T0135968

Diss. ETH Nr. 17423

Thin films on icosahedral AlPdMn quasicrystal

A dissertation submitted to the
ETH Zurich

for the degree of
Doctor of Sciences ETH Zürich

presented by

JEAN-NICOLAS LONGCHAMP
Dipl. Phys. ETH
born on January 22, 1980
citizen of L'ISLE, VD

accepted on the recommendation of

Prof. Dr. M. Erbudak, examiner
Prof. Dr. D. Pescia, co-examiner
Prof. Dr. W. Steurer, co-examiner

2007

Bibliografische Information der Deutschen Nationalbibliothek

Die Deutsche Nationalbibliothek verzeichnet diese Publikation in der Deutschen Nationalbibliografie; detaillierte bibliografische Daten sind im Internet über http://dnb.d-nb.de abrufbar.

©Copyright Logos Verlag Berlin GmbH 2007
Alle Rechte vorbehalten.

ISBN 978-3-8325-1781-6

Logos Verlag Berlin GmbH
Comeniushof, Gubener Str. 47,
10243 Berlin
Tel.: +49 030 42 85 10 90
Fax: +49 030 42 85 10 92
INTERNET: http://www.logos-verlag.de

Pour mes parents que j'aime

Qui es-tu?

Le monde de Sophie.

Jostein Gaarder, 1952–?

Contents

List of Figures

Résumé

Les quasicristaux et particulièrement les interfaces cristal-quasicristal fascinent les scientifiques des surfaces depuis plus de deux décennies à cause des phénomènes inattendus qui apparaissent à la conjonction de structures incompatibles, mais néanmoins, liées. Dans cette dissertation, l'oxydation à haute température de la surface de symétrie d'ordre cinq du quasicristal icosaèdre $Al_{70}Pd_{20}Mn_{10}$ est discutée en premier lieu. La formation d'un film bien ordonné d'Aluminium-oxyde d'une épaisseur de 5 Å représente une toute nouvelle classe d'interface cristal-quasicristal. La composition chimique de la surface est mesurée grâce à la spectroscopie des électrons d'Auger et par la spectroscopie des photoélectrons qui confirment qu'uniquement les atomes d'Aluminium du substrat sont oxydés. Les motifs obtenus des films Aluminium-oxyde, grâce à la diffraction d'électrons lents, partagent beaucoup de similarités avec ceux obtenus après l'oxydation à haute température des faces NiAl(110). Cet alliage intermétallique possède la structure CsCl, qui est en forte relation avec la structure icosaèdrale. La face (110) est alors "l'analogue" de la surface de symétrie d'ordre cinq du quasicristal. L'affinité réciproque des deux structures est illustrée par la présence de domaine AlPd de type CsCl à la surface du quasicristal après bombardement de celle-ci avec des ions Ar^+. Le film d'Aluminium-oxyde consiste, comme dans le cas de NiAl(110), de paires de domaines gamma-Al_2O_3. A cause de la symétrie d'ordre cinq, cinq paires de ces domaines séparées par 72° sont observées dans notre cas. Chaque paire est alignée le long d'une des cinq directions d'ordre deux présentes à la surface du quasicristal. Cette orientation est la conséquence de l'affinité entre les structures CsCl et icosaèdrale. Dans les motifs de diffractions, des reflexes provenant du quasicristal peuvent être observés après le processus d'oxydation. Ceci nous montre que l'ordre quasi-périodique n'est pas dérangé par l'oxydation. La taille moyenne des domaines d'Aluminium-oxyde a été évaluée à environ 35 Å.

Les films d'oxydes sont d'importants supports pour la catalyse métallique et les dispositifs magnéto-électroniques. Dans le premier cas, un composant catalytiquement actif, par exemple un métal de transition, est dispersé sur un support, normalement un oxyde tel que silicate ou alumina. De plus, les films d'aluminium oxyde sont, à cause de leur caractéristiques électriques et de leur remarquable absence de relief, utilisés largement comme support pour l'épitaxie de matériaux semi-conducteurs. PbTe et CdTe sont deux semi-conducteurs binaires industriellement importants. Le premier est un semi-conducteur IV-VI avec une petite bande interdite utilisé comme détecteur infrarouge, pendant que le deuxième, un semi-conducteur II-VI , est un des plus importants matériaux pour les cellules photovoltaques. Dans cette dissertations, nous avons utilisé la surface oxydée d'ordre cinq de AlPdMn comme substrat pour la déposition de PbTe et CdTe. Les motifs de diffractions obtenus pour des films des deux matériaux consistent, au lieu des points habituels, de cercles. Ils sont caractéristiques de nano-cristaux possédant une orientation azimutale aléatoire mais une orientation polaire bien définie; la face (001) et la face (111) dans le cas de PbTe et CdTe respectivement. Grâce aux motifs de diffraction une taille moyenne des domaines de 35 Å a pu être évaluée, qui correspond à la taille des domaines d'Aluminium-oxyde. Nous argumentons qu'au contraire du cas normal en épitaxie où la taille des domaines est donnée par la différence des paramètres de réseaux, celle-ci est donnée dans notre cas par la taille du substrat. Ceci est confirmé par les résultats obtenu lors de la déposition de Al sur le même substrat. Dans ce cas, nous observons des domaines possédant une structure cubique-face-centrée avec leur face (111) parallèle au substrat et une taille moyenne similaire.

La spectroscopie des photoélectrons avec résolution angulaire a été utilisée sur les films de PbTe. Pas d'effets provenant du confinement des électrons n'ont pu être mesuré. Pour que cette mesure ait pu avoir lieu, nous avons dû exposer ce film à l'air. Nous pensons que cette procédure a changé les propriétés électriques du film, nous empêchant d'observer tout effet de taille. Nous présentons aussi des mesures de spectroscopie des photoélectrons avec résolution angulaire sur des films de Ag croît sur la surface de symétrie d'ordre cinq de AlPdMn et sur la surface de symétrie d'ordre dix de AlCoNi comme modèle d'effets de confinement apparaissant à cause des symétries incompatibles du film et du substrat. Ag, dans une structure cubique-face-centrée, expose dans les deux cas leur face (111) parallèle à la surface des substrats. En analysant les états puits quantique *sp*, nous affirmons que l'interface avec le quasicristal constitue une barrière efficace contre la propagation des électrons, dû à l'absence de symétrie commune entre les fonctions d'onde de type Bloch et critiques.

Finalement, nous présentons les résultats de la déposition de Si et Ge sur la surface de symétrie d'ordre cinq de AlPdMn. Grâce à la diffraction d'électrons lents et l'imagerie d'électrons secondaires, nous avons pu déterminer que les deux matériaux croissent de manière trois dimensionnelles et amorphes. Ceci est la conséquence de la haute directionnalité des liaisons atomiques. A des températures au-dessus de 370 K, Si diffuse dans le substrat, ce qui nous empêche d'observer un probable mode de croissance ordonné à haute température. L'effet de l'incorporation de Si dans le quasicristal sur la température de Debye de surface est présenté en dernier lieu. Une croissance linéaire de celle-ci en fonction du montant de matériau incorporé a été mesurée. Nous suggérons que la diffusion de Si dans AlPdMn est un processus supporté par le remplissage des places libre dans le quasicristal.

Abstract

Quasicrystals and particularly crystal-quasicrystal interfaces are fascinating surface scientists for more than two decades due to the several unexpected phenomena that occur at the conjunction of incompatible, but nevertheless related structural properties. In this dissertation, the oxidation at high temperature of the fivefold-symmetry surface of an icosahedral $Al_{70}Pd_{20}Mn_{10}$ quasicrystal is first discussed. The formation of a 5 Å thin well-ordered aluminum-oxide film represents a complete new kind of crystal-quasicrystal interface. The stoichiometry of the near-surface region is investigated by means of Auger electron spectroscopy and x-ray photoelectron spectroscopy and both confirm the oxidation of only the Al atoms of the quasicrystalline substrate. The patterns obtained by means of low-energy electron diffraction from the aluminum-oxide share several similarities with those obtained from the oxidation at high temperature of NiAl(110) faces. This intermetallic alloy posses the CsCl structure, which has a strong relationship with the icosahedral structure. The (110) face is then the "analogue" of the pentagonal surface of an icosahedral quasicrystal. The affinity of the two structures is illustrated by the CsCl-like AlPd domains observed, by means of secondary-electron imaging, after Ar^+-sputtering of the quasicrystalline surface. The aluminum-oxide film consists, as in case of NiAl(110), of pairs of γ-Al_2O_3-like domains. Due to the fivefold symmetry of the quasicrystalline surface, five pairs of these domains rotated by 72° with respect to each other are observed in our case. Each pair is aligned along one of the five twofold-symmetry directions present in the fivefold-symmetry surface and is a consequence of the strong affinity of the icosahedral structure for the CsCl structure. In the diffraction patterns, spots arising from the quasiperiodic substrate are after the oxidation process still present, pointing out that the quasicrystalline order is not disturbed by the oxidation process. The average in-plane size of the aluminum-oxide domains was evaluated to be approximately 35 Å.

Oxide thin films are important supports for dispersed metal catalysts and magnetoelectronic devices. In the former, for instance, a catalytically active component such as a transition metal is dispersed over a suitable support material, usually an oxide like alumina or silica. Aluminum-oxide films are, because of their electronic characteristics and their remarkable flatness, also widely used as substrate for semiconductor epitaxy. PbTe and CdTe are two important binary semiconductors. The former is IV-VI narrow band-gap semiconductor used in infrared detectors, while the latter, a II-VI semiconductor exhibiting a wide-band gap, is one of the most important material in the photovoltaic cells research field. In this dissertation, we used the oxidized fivefold-symmetry surface of i-AlPdMn as substrate for the deposition of PbTe and CdTe. Diffraction patterns obtained from thin films of both materials exhibit, instead of the usual spots, diffraction rings. They are characteristics of nanocrystallites having a random azimuthal orientations but a well-defined polar orientation; the (001) face and the (111) face in case of PbTe and CdTe, respectively. From the diffraction patterns, average domain sizes of 35 Å were deduced, which corresponds to the average size of the aluminum-oxide domains. We argue that, in contrast to normal epitaxy where the domain size of the deposited material is given by the lattice mismatch between the growing film and the substrate, it is here given by the size of the substrate. This was confirmed by the results obtained from the deposition of Al onto the same substrate. Face-centered-cubic Al(111) domains with a similar average size are observed in this case.

Angle-resolved photoemission spectroscopy investigations on the PbTe films have been performed. No significant confinement effects could be observed. Because the samples have been exposed to air, in order to transfer them to the measurement chamber, we argue that oxygen has changed the electrical properties of the films, preventing us to observe any size effect.

We also present angle-resolved photoemission spectroscopy measurements on Ag films deposited onto the fivefold-symmetry surface of icosahedral AlPdMn and onto the tenfold-symmetry surface of decagonal AlCoNi as model for confinement effects occurring due to the incompatible symmetries between the crystalline films and the quasicrystalline surfaces. Ag, in a face-centered-cubic structure, exhibit in both cases its (111) face parallel to the substrates surface. By analyzing the Ag *sp*-derived quantum-well states, we assert that the interface with the quasiperiodic material constitutes an efficient barrier for electron propagation, due to lack of common point-group symmetries between Bloch-like and critical wave functions.

Finally, the results from the deposition of Si and Ge onto the fivefold-symmetry surface of icosahedral AlPdMn are reported. Low-energy electron diffraction and secondary-electron imaging investigations lead to the conclusion that both materials have a three-dimensional amorphous growth mode for a large substrate temperature range. This is, probably, a consequence of the high directionality of the atomic bonds in these semiconductor materials. At temperature above 370 K, Auger electron spectroscopy measurements show that Si noticeably diffuses into the substrate, which prevents a probable crystalline growth mode at high temperature. The effect of the incorporation of Si in a near-surface region on the surface Debye temperature is at last presented. A linear increase of the latter as function of the incorporated amount of material was measured. We suggest that Si diffusion into icosahedral AlPdMn is a vacancy-mediated process.

Chapter 1

Introduction

In his famous talk of 1959, Richard P. Feynman, pointed out that there is "plenty of room at the bottom" [1]. He predicted the existence of exciting new phenomena at the atomic scale which could possibly revolutionize science and technology. On one hand, experimental techniques like, for instance, atomic force microscope [2] or scanning tunneling microscope [3] have been since then developed, allowing us to image and manipulate molecules and atoms. On the other hand, we now have access to nanostructured material, such as semiconductor quantum dots and carbon nanotubes, which exhibit extraordinary properties. Both of these developments are parts of the nanotechnology, whose goal is to be able to produce and use devices with sizes in a range 1-100 nm.

In the late 60's and the early 70's, the first microprocessors were brought on the market. The 4-bit chips comprised approximately 4000 transistors [4]. Nowadays, the commercial microprocessors contained more than one billion transistors per chip. The density growth appears to follow the famous Moore's law which affirms that the number of transistors on an integrated circuit for minimum component cost doubles every 24 months [5]. A similar exponential growth trend exists for the magnetic storage devices, exhibiting an even faster pace [6, 7]. In order to follow these trends, industry needs ever and ever smaller reliable devices whose sizes are gradually approaching the nanometer range. Therefore, nanotechnology might give solutions to the industrial aspiration of miniaturization. But, even if reliable nanodevices possessing suitable characteristics become available, mass fabrication technics of such elements have to be developed in order to use them in an industrial basis.

There are two fundamental strategies for the creation of well-ordered nanostructures on substrates in controlled and reproducible manner, the "top-down" and "bottom-up" techniques (Fig. 1.1). With the top-down approach, which mainly consists of the electron-beam writing and nanolithography techniques, one is capable to produce structures with sizes around 100 nm. Electron-beam writing achieves very good spatial resolution [8] but is not an applicable method for production in series. Lithography at the nanoscale, which uses x-ray radiation, mainly ultraviolet [9], will need tremendous upgrade in the industrial equipments, such as optics and radiation sources, in order to achieve mass production of nanostructures.

The common principle to all top-down techniques is that they basically impose a structure on the substrate. In contrast, bottom-up techniques take advantage of the self-assembly mechanisms which occur if the right combination of adsorbate and substrate is chosen. The growth results from the competition between kinetic and thermodynamic effects. If the diffusion of the adsorbate atoms on the substrate surface is faster than the deposition, the adsorbed species have time to reach a position on substrate which corresponds to an energy minimum. If the deposition is faster than surface diffusion, the growth is mainly governed by kinetics. For instance, low-temperature growth of metal nanostructures on metal surfaces is a prototype of kinetically controlled growth mechanisms. This is mainly explained by the fact that metal bonds have essentially no directionality that can be used to direct interatomic interactions. Consequently, the size and shape of the metal nanostructures are determined by the competition between the different movements that the atoms can make along the surface, such as diffusion of adatoms on surface terraces, over steps, along edges, and across corner or kinks [9, 10]. Ideally their morphology can then be tuned by influencing the kinetic conditions.

In case of semiconductors deposition onto metal surfaces, the situation is quite different because of their extreme directional bonds. Here, thermodynamics play the key role in the growth mechanisms. In the thermodynamical limit, experimental results point to the existence of three different growth modes (Fig. 1.1). In the Frank-Van der Merwe mode, the growth follows a layer-by-layer scenario, the film remaining completely two-dimensional. The Volmer-Weber growth is just the opposite. Three-dimensional crystallites nucleate immediately upon contact and the overlayer may not completely cover the exposed substrate surface until a great number of atoms have been deposited. The Stranski-Krastanov mode is an intermediate case, here a few monolayers adsorb in layer-by-layer mode before three-dimensional structures begin to form [11].

Efforts to produce and characterize semiconductor quantum dots have mainly been driven by the desire to develop systems that take advantage of their extremely

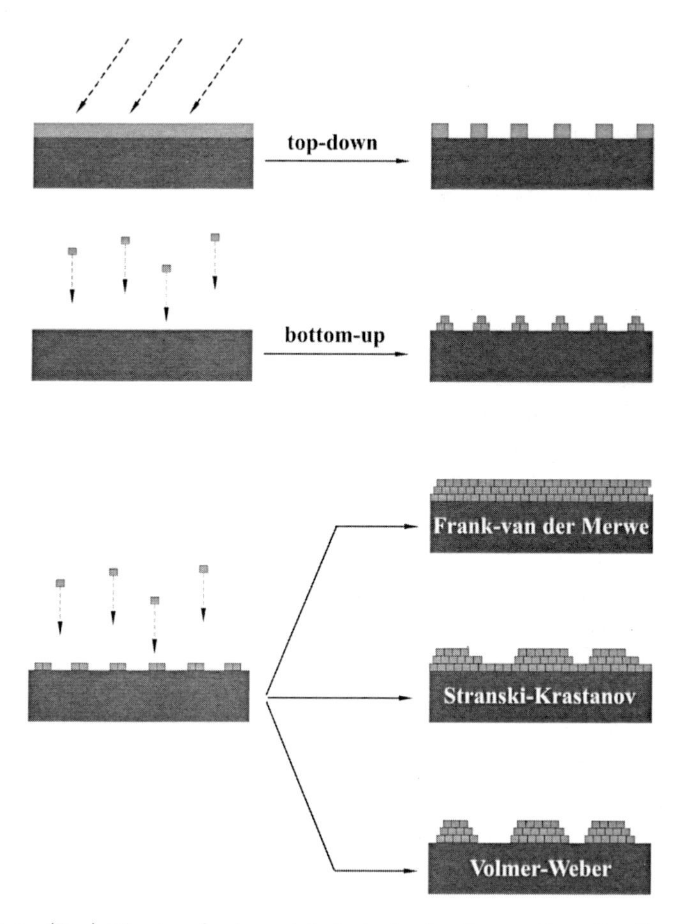

Fig. 1.1: (Top) The two fundamental strategies for the creation of well-ordered nanostructures on substrates. The "top-down" and "bottom-up" approaches. (Bottom) Schematic view of the three topologically distinct epitaxial growth modes.

small size and low power dissipation. For instance, the performance of lasers can be substantially improved by using quantum dots as active medium [12, 13]. The tunable and discrete energy levels typical of such structures mean that the choice of emitted wavelength can be adjusted with an incredible flexibility. The small active volume permits laser operation at low power and high frequency independently of the working temperature. In semiconductor nanocrystals, quantum confinement effects already arise for structures with dimensions of $10 - 100$ nm and containing something like 10^3 to 10^6 atoms in a crystalline lattice. For instance, in the II-VI semiconductor CdS the band-gap energy evolves from 4.5 to 2.5 eV as the size increases from the molecular regime to the macroscopic range [14]. The energy required to add an excess charge carrier above the band gap decreases by 0.5 eV [15] and the melting point increases from 400 to 1600 °C [16]. Because the exact physical properties of the quantum dots depend on the shape and size, it is essential to know their actual morphology.

But even though they were intensely studied for more than a decade and many of their electronic and optical properties characterized, the structures of nucleated semiconductor islands remain incompletely understood. Nevertheless, recently for the growths of Ge on Si(100) and InAs on GaAs(001), only two discrete, well-defined families of three dimensional islands were observed [17, 18], small islands that are bounded by one type of shallow facets and referred to as pyramids, and larger, multi-faceted islands that are characterized by steeper facets and referred to as domes. These experimental observations confirm theoretical predictions [19] that well-defined island shapes occur during growth, independently of the specific material system considered.

Quasicrystals, also called quasi-periodic crystals, form a third state of atomic ordering besides the periodic and amorphous structures. They were discovered by Shechtman et al. [20] in 1984 by observing a tenfold-symmetric diffraction pattern of a rapid quenched AlMn alloy. They also noticed the presence of two- and threefold symmetry axes with the angles between them corresponding to a sample possessing icosahedral point-group symmetry. While the very existence of a diffraction pattern at all indicates long-range order in atomic positions, the presence of the fivefold-symmetry axes, forbidden in crystalline solids, underline the paradox presented by the quasicrystals. Levine et Steinhardt [21], proposed that the translational order in quasicrystals might be quasiperiodic rather than periodic, but allowing long-range order, which makes possible the existence of diffraction patterns. In case of quasicrystals, they consist of a set of Bragg peaks which densely fill the reciprocal space, since there are no translational symmetries, and consequently, there is no minimum spacing between diffraction reflexes [22]. However, in experiments only spots with intensities above a minimum threshold are observed [23].

The most well-known quasiperiodic arrangement may be the Fibonacci sequence, noticed during observation on rabbit reproduction. Consider the following rules for birth and maturation of rabbits. Start with a single pair of mature rabbit (denoted by L for "large"). In each generation every L rabbits gives birth to a new pair of baby rabbits (denoted by S for "small"), while each S rabbits matures into L rabbits. The evolution of the rabbit population can be established as follows. Start with an L and an S side by side along a line. Replace the L with LS and the S with L to obtain LSL and repeat the procedure for every new generation. The number of pair of rabbits present after each generation are the Fibonacci numbers. The population grows exponentially over time, with the population of each generation approaching τ, the golden mean, multiplied by the population of the previous generation [24]. The sequence of L and S forms a quasiperiodic order and is an one-dimensional equivalent of a quasicrystal atomic arrangements. Examples of equivalent two-dimensional quasiperiodic ordering are the famous Penrose tilings [21]. Besides the fivefold-symmetry, quasicrystals exhibit other crystallographically forbidden symmetries, such as, and so far observed, eight-, ten-, and twelvefold symmetries.

Quasicrystals are all binary, ternary, or quaternary alloys whose great majority is Al based [23]. Two noticeable examples of stable quasicrystalline compounds are the decagonal (d-)AlCoNi and the icosahedral (i-)AlPdMn. Quasicrystals can be grown slowly and carefully using techniques for growth of high-quality conventional crystals [25]. The alloy stoichiometries play a key role in the existence of quasicrystalline phases, for instance, the icosahedral phase of AlPdMn exists only if the initial composition of melt, from which the quasicrystal is formed, is in the narrow range 71 − 78 at.% Al, 15 − 22 at.% Pd, and 4 − 10 at.% Mn [26]. The origin of the existence of stable quasicrystalline phases remains in question. No proven explanation clarifies why a material favors crystallographically forbidden rotational symmetry and translational quasiperiodicity when at nearby chemical compositions it forms more conventional crystal structures. Besides their extraordinary structures and symmetries, quasicrystals possess unusual physical properties. Their elastic properties are characterized by the presence of phasons, which corresponds to rearrangements of the relative atomic positions. Unlike their constituent elements, which are normally good electrical conductors, quasicrystals conduct electricity poorly. Their resistivity grows with the perfection of their quasicrystalline order. This behavior is consistent with the observation of a gap in the electronic density of states at the Fermi level. Such gap, also called pseudo-gap, may also play a role in the stability of the quasicrystalline phases. Since the electrons present at the Fermi level are those with the highest energy, diminishing their number may lower the overall energy, this is known as the Hume-Rothery rule [27]. Quasicrystals also possess other unusual physical properties as high hardness [28], low surface free energy [29], and low thermal conductivity [30].

Crystal-quasicrystal interfaces are of special interest due to the several unexpected phenomena that occur at the conjunction of different, but nevertheless related structural properties. Such interfaces can be produced in three different ways. The first one is the deposition of quasicrystalline materials on crystalline substrates. For instance, AlCoNi [31] and AlCuFeCr [32] films grown on sapphire(0001) consist of quasicrystalline domains, whose orientations are mediated by the crystalline substrate. The second art of crystal-quasicrystal interfaces is produced by ion sputtering of quasicrystalline surfaces. Ion bombardment of alloys surfaces induces, due to preferential sputtering of the light atoms, a change in the chemical composition in a near surface region. In the case of i-AlPdMn, Ar^+ bombardment of the fivefold-symmetry surface results in the formation of CsCl-type $Al_{50}Pd_{50}$ domains exposing either their (110) or their (113) faces parallel to the surface. Depending on the sputtering conditions such as the substrate temperature or the incidence angle of the ions, single domains overlayers as well as overlayers consisting of five domains azimuthally separated by 72° were observed for both polar orientations. The third method to produce crystal-quasicrystal interfaces is to deposit crystalline materials on quasicrystalline surfaces. During the last decade, a great number of adsorbates were used in this purpose (for a complete review see [33]). Besides heterogeneous nucleation and pseudomorphic growth, the growth of crystalline self-size-selecting crystalline domains with dimensions in the nm range and orientations mediated by the substrates has been observed.

Particularly interesting is the case of the deposition of Co onto the fivefold-symmetry surface of i-AlPdMn [34]. For submonolayer Co deposits, an atomic layer consisting of five AlCo domains with nm dimensions is formed at the surface. The CsCl-type domains expose their (110) face parallel to the pentagonal surface and are rotated by 72° with respect to each other. For further deposition, Co in the body-centered cubic structure grows epitaxially on the AlCo domains. The orientational relationship between the CsCl-type AlCo domains and substrate is, in this case, explained by the optimum matching of the average structures [33]. Steurer previously explained the orientation of sputter-induced CsCl-type domains on i-quasicrystals [35,36] by the optimum matching of the average structures. Furthermore, Dmitrienko and Astaf'ev have revealed the close relationship between the i-quasicrystal and the CsCl structure by introducing a three-dimensional model for the growth of i-quasicrystalline structures [37].

The structure of this dissertation is as follows. Chapter 2 deals with the experimental procedures used in this work. Chapter 3 presents the results of the exposure of the pentagonal surface of i-AlPdMn to O_2 at high temperature and the orientational relationship of the oxide film towards the quasicrystalline surface. In Chapter 4, the results of the deposition of PbTe and CdTe, two direct band-gap semiconductors, onto the oxidized fivefold-symmetry surface of i-AlPdMn. This Chapter also deals with the size relationship between the aluminum-oxide domains present at the surface and the size of the semiconductor domains. Angle-resolved photoemission measurements performed on these films are presented. Confinement effects arising in Ag films deposited onto the five-fold symmetry surface of i-AlPdMn are reported at the end of this Chapter. In Chapter 5, the reader can find the results of the deposition of Si and Ge onto the pentagonal surface of i-AlPdMn and the evolution of its Debye temperature as function of the incorporated amount of Si. Chapter 6 concludes this experimental work and ends with an outlook on the possible applications of the presented results.

Chapter 2

Experimental setup and methods

2.1 The quasicrystalline sample

A single-grain i-quasicrystal with a nominal bulk composition of $Al_{70}Pd_{20}Mn_{10}$ and grown by the Bridgman technique [38] was oriented by the x-ray Laue [39] method along a fivefold-symmetry axis and cut by spark erosion perpendicular to this direction [40]. After polishing the sample using diamond pastes with grain sizes down to 0.5 μm, the sample with dimensions of $7 \times 4 \times 1$ mm^3 was mounted on a goniometer and introduced into an ultra-high vacuum (UHV) chamber with a base pressure in the lower 10^{-10}-mbar range. The sample temperature was measured with a chromel-alumel (K-type) thermocouple [41] pressed onto the front surface and the sample was heated from the backside using a resistance heater. The quasicrystalline order at the surface was established by cycles of Ar^+-ion sputtering at 1.5 keV and subsequent annealing at 700 K for 1 hour. This preparation procedure leads to a bulk-terminated surface [42]. The specimen could be cooled down, to minimum temperature of 200 K, using a Cu cooling feed which was brought in contact with liquid N_2.

2.2 Surface structure

The atomic surface structure was investigated by means of low-energy electron diffraction (LEED) and secondary-electron imaging (SEI). A surface sensitive electron diffraction experiment requires electrons with a wavelength comparable with the typical interatomic spacing of approximately 1 Å and a minimal penetration depth in the sample. Fortunately, electrons with kinetic energy in the range

$20 - 500$ eV possess both of these characteristics and are therefore suitable for such an experiment. LEED patterns are images of the reciprocal lattice of the surface structure and give information on the long-range order and defects present at the surface [43]. In case of SEI, electrons with primary energy of 2 keV and a penetration depth of approximately $20 - 50$ Å were used. In this energy regime, primary electrons scatter at atoms and excite secondary electrons which are focused along densely packed atomic rows due to interaction with the Coulomb potential of the neighboring atoms [44]. Furthermore, the secondary electrons can interfere between dense atomic planes, giving rise to the so-called Kikuchi bands [45]. The schematic experimental setup for LEED and SEI is displayed in Fig. 2.1. It consists of a three-grid back-view display system, with a total opening angle of 100° and operated with a beam current in the low μA range. All the patterns were recorded using a 16-bit charge-coupled device camera and were subsequently normalized by the overall response function of the display system in order to eliminate spurious signals [46].

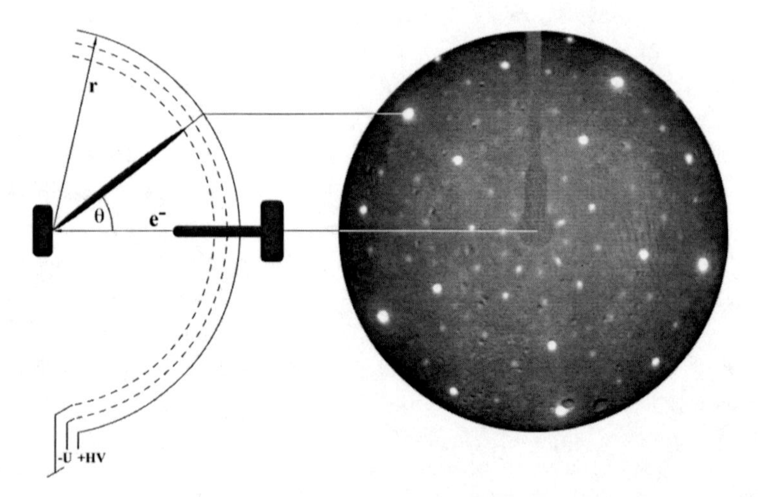

Fig. 2.1: Schematic illustration of the LEED and SEI setup. The diffraction pattern presented was obtained from the fivefold-symmetry surface of i-AlPdMn at a primary-electron energy of 55eV.

Spot-size analyses have been performed on several diffraction patterns in order to evaluate the domain sizes. As example, Fig. 2.2 presents a line scan of the normalized diffraction intensity after deposition of PbTe onto the oxidized fivefold-symmetry surface of i-AlPdMn. The profiles were analyzed according to the method proposed by Horn-von Hoegen [47]. In this figure, the radial intensity profile of the diffraction spots at a distance in the reciprocal room $q \approx 0.42$ Å^{-1}, i.e., at a polar angle $\theta \approx 36°$ for a primary energy of 90 eV, is presented as example for the evaluation of the domain sizes using LEED patterns. The full width at half maximum value of the diffraction intensity is $\Delta k \approx 0.029$ Å^{-1}, resulting in an average domain size of approximately 35 Å.

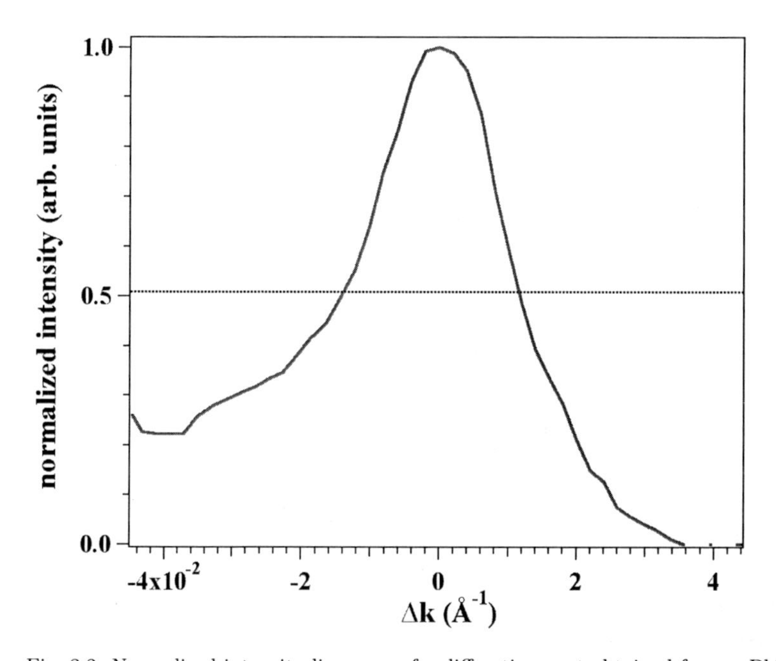

Fig. 2.2: Normalized intensity line scan of a diffraction spot obtained from a PbTe film grown on the oxidized pentagonal surface of i-AlPdMn. This particular case is here presented as example for the evaluation of the domain sizes with LEED experiment.

2.3 Electronic structure

Electronic structure investigations on the quasicrystalline sample and the grown films were performed in four different ways, Auger-electron spectroscopy (AES), x-ray electron spectroscopy (XPS), angle-resolved photoelectron spectroscopy (ARPES) using synchrotron radiation, and ARPES using He Iα radiation.

AES measurements were performed to determine the chemical composition of a near surface region. Electrons of energy $2-20$ keV are incident upon a conducting sample. These electrons cause core electrons from atoms contained in the sample to be ejected resulting in an atom with a core hole. The atom then relaxes via electrons with a lower binding energy dropping into the core hole. The energy thus released can be converted into an x-ray or emit an electron. This electron is called an Auger electron after Pierre Auger who discovered this relaxation process. After the emission of the Auger electron, the atom is left in a doubly ionized state. The energy of the Auger electron is characteristic of the element that emitted it, and can thus be used to identify the element. The short inelastic mean free path of Auger electrons in solids ensures the surface sensitivity of AES.

XPS which is based, as ARPES, on Hertz's photoelectric effect, was used in order to investigate the density of states of the electrons at the Fermi level and of the core level electrons in a near surface region. The photoionization of the sample was performed by non-monochromatized AlK$_\alpha$ (1486.6 eV) and MgK$_\alpha$ (1253.6 eV) sources and the spectra were recorded by a cylindrical mirror energy analyzer equipped with a postmonochromator and the total energy resolution was 1 eV.

Photoelectron spectroscopy measurements on Ag films grown on the fivefold-symmetry surface of i-AlPdMn and the tenfold-symmetry surface of d-AlCoNi (see 4.3) were performed at VUV beamline of the Elletra synchrotron facility in Trieste, Italy. In this case, the photon energy was chosen to be 40 eV, which is the best compromise between cross section, surface sensitivity, and energy resolution. The latter was typically several meV. The investigations on a PbTe film grown on the oxidized pentagonal surface of i-AlPdMn were performed at the ARPES facility of the Surface Physics Group of the University Zurich. These experiments have been performed in a VG ESCALAB 220 spectrometer, modified with a computer controlled two-axis sample goniometer [48]. All data were taken at room temperature using He Iα radiation. The energy/angle resolution was set to 40 meV/1°.

2.4 Molecular-beam epitaxy

Si, Ge, PbTe, and CdTe of purity greater than 99.99% were evaporated using power-regulated molecular-beam sources, while the sample was kept at constant temperature. After the initial outgassing procedure of the sources, the pressure never exceeded 1×10^{-9} mbar during evaporation. The deposition rates were calibrated in a separate experiment by measuring the 92eV $L_{2,3}VV$, the 1147 eV $L_3M_{4,5}N_{4,5}$, the 94 eV NOO, the 376 eV $M_5N_{4,5}N_{4,5}$, and 483 eV $M_5N_{4,5}N_{4,5}$ AES signals of Si, Ge, Pb, Cd, and Te, respectively, during deposition onto a polycrystalline Cu sample. The deposition rates were found to be for Si 0.7 ± 0.05, Ge 0.8 ± 0.05, PbTe 1.05 ± 0.05 and for CdTe 1.41 ± 0.05 Å/minute. The values of the inelastic mean free paths of the measured electrons, used in the calibration procedure, were calculated according to the model of Tanuma et al. [49].

During the oxidation steps, O_2, with a purity of 99.998%, partial pressure was maintained between 5×10^{-7} and 1×10^{-8} mbar. The sample was kept at constant temperature and, after the evacuation of the oxygen out of the chamber, annealed at 700 K. Typical exposure times were 20 minutes for low adsorption, 4 hours for high adsorption and the annealing process was performed during 1 hour. The exposure is given, in the following, in L units (1 L= 1.3×10^{-6} mbar·s).

Chapter 3

Oxidation of the pentagonal surface of an icosahedral AlPdMn quasicrystal

3.1 Oxidation of binary X-Al alloy surfaces at high-temperature

Metal oxide interfaces are of considerable interest since they are, among other applications, important supports for dispersed metal catalysts and magnetoelectronic devices [50,51]. In the former, for instance, a catalytically active component such as a transition metal is dispersed over a suitable support material, usually an oxide like alumina or silica [52]. In the first place, this is done in order to achieve the highest possible surface area of the active phase. Because of the high degree of dispersion, however, particle-size effects, originating from specific structural or electronic features and metal substrate interactions can influence the catalytic behavior significantly. The most obvious choice of oxide substrates are single crystal samples, as Al_2O_3, of course. When studying bulk oxides, however, some problems arise. Their insulating character forbids electron and ion spectroscopy, as well as LEED and scanning tunneling microscope (STM) measurements. Other difficulties arise from the poor thermal conductivity of the bulk oxides and from their problematical surface cleaning. Therefore, ultra-thin oxide films grown on a metallic substrate are an excellent alternative in order to overcome all these problems [53,54]. It has been shown that already films with a thickness of just a few Å can exhibit physical properties of the bulk material [55]. Such films can be prepared by oxidation of metal single crystal, although this leads, in most of the cases, to the formation of an amorphous overlayer. This is, for instance, the case for the oxidation of aluminum single crystals [56].

Al_2O_3 phase	Crystal system	$a, b, c(\text{Å}), \beta$	Oxygen sub-lattice
α-Al_2O_3	Hexagonal	$a = 4.759$	hcp
		$c = 12.99$	
κ-Al_2O_3	Hexagonal	$a = 9.71$	
		$c = 17.86$	
γ-Al_2O_3	Cubic	$a = 7.91$	fcc
γ'-Al_2O_3	Distorted-Cubic	$a_1 = 10.60$	Distorted-fcc
		$a_2 = 17.90$	
		$\alpha = 88.7°$	
θ-Al_2O_3	Monoclinic	$a = 5.64$	fcc
		$b = 2.92$	
		$\beta = 104°$	
δ-Al_2O_3	Tetragonal	$a = 7.96$	fcc
		$c = 11.70$	
β-Al_2O_3	Hexagonal	$a = 5.60$	
		$c = 22.5$	
a-Al_2O_3	–	–	Dense random packing

Table 3.1: Crystal structure of some important aluminum oxides [57].

On the contrary, after oxidation at high temperatures, surfaces of alloys such as NiAl, Ni_3Al, CoAl, or FeAl consist of well-ordered aluminum-oxide (Al-ox) films [51]. For instance, oxidation of the CoAl(100) face leads to the formation of a θ-Al_2O_3 film [58] (for a comparison of the different Al_2O_3 phase see table 3.1). Exposure of the $Ni_3Al(111)$ surface to O_2 results in a Al_2O_3 film with a complicated superstructure $\sqrt{67} \times \sqrt{67}\ R47.784°$ [59]. Alumina films grown on Ni-Al alloys exhibit a high degree of crystallinity, a very low surface roughness and an excellent reproducibility in preparation [60]. Consequently, the oxidation at high temperature of Ni-Al crystal surfaces was chosen as model for the selective oxidation of intermetallic alloys.

NiAl possesses the CsCl structure with a lattice constant of $\sim$ 2.89 Å. Oxygen exposure of a NiAl(110) face at high temperature leads to the formation of a well-ordered Al-ox film. Different but nevertheless related models are proposed for the structure of this film. Lykhach et al. base their model on the analyses of the LEED patterns obtained from the grown film. They propose a crystal structure comparable to γ-Al_2O_3, one of the transition phases of alumina used as a support in technical catalysis. The oxide overlayer was found to be 5 Å thick, if the substrate surface is exposed to oxygen until saturation, and oxygen terminated [61]. The unit cell of the terminating oxygen layer of the modified γ-phase, has $a_1 = 10.6$ Å, $a_2 = 17.9$ Å and $\alpha = 88.7°$ as lattice parameters. The γ-film has its nominal, since slightly distorted, (111) plane parallel to the NiAl(110) substrate surface and consists of domains with two possible azimuthal orientations determined by the alignment of the elementary oxygen cells along the $[1\bar{1}\bar{1}]$ and $[1\bar{1}1]$ directions of NiAl(110). Consequently, the angle between the $[1\bar{1}0]$ direction of NiAl and both the $[01\bar{1}]$ and $[10\bar{1}]$ directions of the oxide domains is $\pm 24°$ [62].

Kresse et al. propose another model based on STM measurements performed on similar Al-ox films [63]. Contrary to the previously presented model, they argue that the exact stoichiometry of the film is not Al_2O_3 but $Al_{10}O_{13}$ [more precisely $4(Al_4O_6Al_6O_7)$]. Oxygen atoms in the surface layers adopt a nearly hexagonal pattern. The unit cell corresponding to this film can be expressed by $a_1 = 10.93$ Å, $a_2 = 17.9$ Å and $\alpha = 91.84°$, which is almost identical to that proposed by Lykhach et al. for the terminating O layer (see above). The difference between the two models arise only from very small atomic displacement within the unit cell ($\sim$ 0.3 Å). Furthermore, Kresse et al. propose a model for the interface NiAl-Al-ox with a preferential position of the Al atoms in the oxide film on top of Ni atoms of the substrate. As Stierle et al., they propose that the film is in one direction commensurate and in the other incommensurate to the substrate.

As mentioned in Chapter 1, crystal-quasicrystal interfaces can also be generated by means of sputtering of quasicrystalline substrates. An ion-bombardment-induced modification of the stoichiometry in a near-surface region destabilizes the quasicrystal structure. In case of i-AlPdMn, sputtering of the fivefold-symmetry surface with Ar^+ leads to the formation of CsCl-type AlPd domains. Two possible orientations of the crystalline domains, depending on the sputtering conditions (temperature, angle of incidence, pre-deposits) are generated. Domains are observed which expose their (110) as well as (113) faces parallel to the pentagonal surface. Additionally, both single-domain overlayers as well as overlayers consisting of five domains rotated by 72° with respect to each other are found [34]. Figure 3.1 displays SEI patterns obtained in cases of (a) five (110) and (b) five (113) domains. These AlPd domains give rise to characteristic LEED patterns (not shown here), whose analyses lead to a lattice constant for the sputter-induced AlPd domains of 2.88 Å. The orientation of the sputter-induced CsCl-type domains on i-quasicrystals was explained by Steurer [35] by the optimum matching of the average structures. The above example of crystal-quasicrystal interface and the results of the growth of Fe, Ni, and Co on the fivefold-symmetry surface of i-AlPdMn [33] demonstrate that there is a close relationship between the icosahedral and CsCl structures (Fig. 3.2).

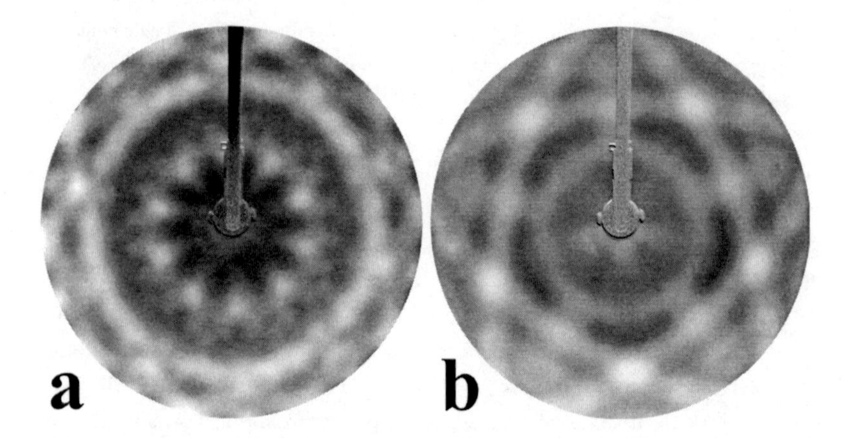

Fig. 3.1: After Ar^+ bombardment of the pentagonal surface of i-AlPdMn (a) five (113) and (b) five (110) domains rotated by 72° with respect to each other can be observed.

Chang et al. and Popović et al. have reported the oxidation of the fivefold-symmetry surface of i-AlPdMn for substrate temperatures between 105 and 870 K [64, 65]. Their XPS investigations on the oxide layers have shown that only the Al atoms of the quasicrystal bind to the oxygen while Pd and Mn remain inert [66]. Contrary to the results obtained in this work, they observe, by means of LEED, the formation of amorphous alumina layers for different oxidation conditions, such as oxygen partial pressure, exposure time and substrate temperature.

3.2 The clean pentagonal surface

Figure 3.3(a) shows the LEED pattern of the pentagonal surface of i-AlPdMn obtained after cycles of Ar^+ sputtering and annealing at 700 K. The quality of the pattern indicates a well-established quasicrystalline order present at the surface [43]. Basis vectors e_i are plotted in the pattern according to those introduced by Schaub et al. [67]. Additionally, the five twofold-symmetry axes lying in the pentagonal surface are displayed. These axes are parallel to $[1\bar{1}0]$ directions of (110) AlPd domains which are obtained by the sputtering of the pentagonal surface. Figure 3.3 (b) displays the corresponding SEI pattern. No traces of crystalline AlPd domains present at the surface, generated by sputtering, could be observed by means of LEED and SEI after annealing.

3.3 Oxidation of the pentagonal surface at high-temperature

The LEED pattern displayed in Fig. 3.3(c) was recorded after exposing a freshly prepared quasicrystalline surface to 800 L O_2 at 700 K and subsequent annealing at the same temperature for 2 hours. Ten peaks originating from the quasicrystalline substrate are still observable (emphasized by the dashed line). The presence of these spots suggests that despite the formation of an oxide overlayer, the quasicrystalline order is conserved by the segregation of Al atoms towards the surface [68]. It also indicates that the thickness of the film is comparable to that reported on NiAl which is 5 Å [60]. In the diffraction pattern, thirty very bright spots are located on a circle. Actually, all the spots originating from the oxide film form concentrical elements of 30, 20, or 10 spots. A comparable density of reflections in LEED patterns were already observed for films generated by the oxidation of intermetallic alloys such as NiAl at a high substrate temperature [61]. Thel rotational symmetry of the pattern is consistent with the existence of multiple domains with different, but well-defined azimuthal orientations. Figure 3.3(d) was obtained from the same surface as for (c) but with the sample holder tilted by 15° in order to reveal the higher order reflections. Spot-size analyses suggests an average domain diameter of $\sim$ 35 Å. Diffraction spots characteristic of the oxide layer, albeit inferior in quality, already emerge after a brief exposure of the clean surface to O_2 at elevated temperatures without postannealing.

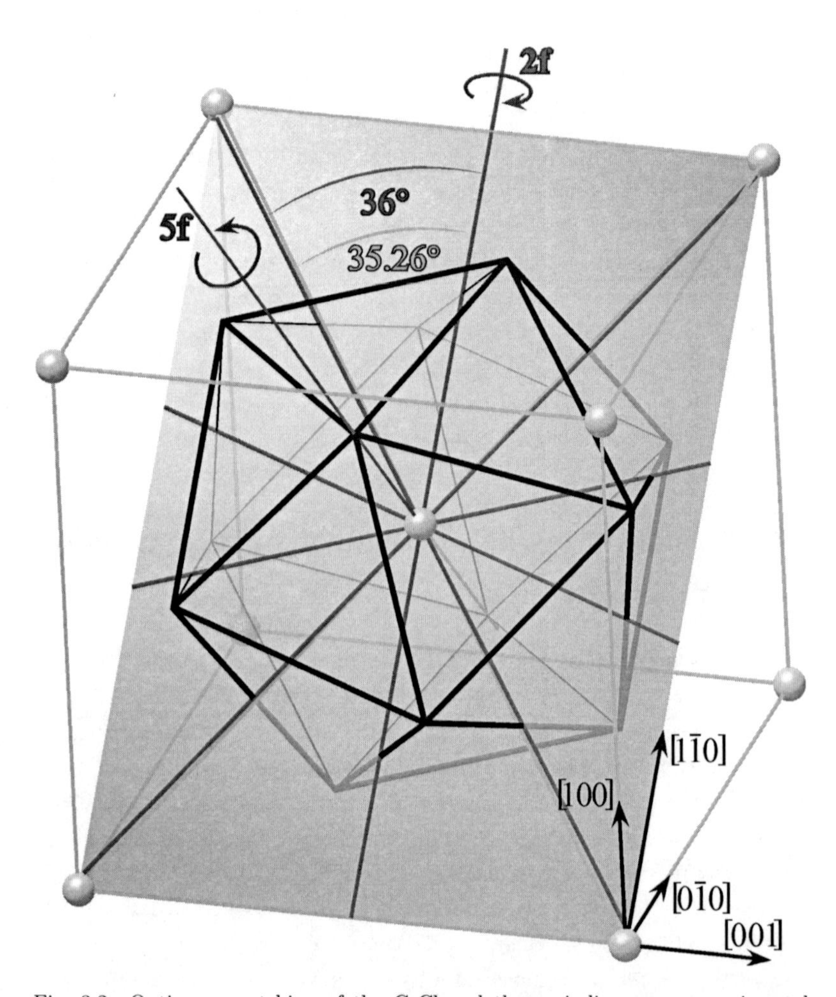

Fig. 3.2: Optimum matching of the CsCl and the periodic average quasicrystal structure. The orientation of CsCl-type domains on icosahedral quasicrystals can be explained by the optimum matching of the CsCl structure with the periodic average quasicrystal structure. In the case of a quasicrystal possessing icosahedral symmetry, represented by an icosahedron, optimum matching requires that the [110] direction of the CsCl-type structure is parallel to a fivefold-symmetry axis of the quasicrystal and the [1$\bar{1}$0] direction is parallel to a twofold-symmetry axis. The angle between two adjacent twofold-symmetry axes is 36°. The angles between the [1$\bar{1}\bar{1}$] as well as the [1$\bar{1}$1] direction and the [1$\bar{1}$0] direction is $\sim 35.26°$.

The presence of only one (00) reflection excludes a faceted structure of the growing film. A multi-domain structure is confirmed by the LEED pattern presented in Fig. 3.3(e) which was recorded after a brief ion sputtering, at an angle of 50° off the normal, of the surface giving rise to the pattern presented in Fig. 3.3(c) and re-exposure to 300 L O_2 at 700 K and subsequent annealing. By this treatment, one azimuthal orientation of the domains is preferentially selected, because 12 of the 30 bright spots observed in pattern (c) are considerably brighter in pattern (e) compared to the remaining 18. The pattern is twofold symmetric and shows close similarities to that obtained after the oxidation of a NiAl(110) surface [69]. One of the two twofold directions of the new structure, indicated in Fig. 3.3(e), is aligned with one of the twofold directions of the substrate. We have applied this sputtering procedure to identify one of the five domains. Indeed, one orientation preferentially survives sputtering better than others. After sputtering, the sample is well annealed at the standard temperature of preparation and LEED spots characteristic of the quasicrystal surface are discernible through the oxide layer. The pattern shown in Fig. 3.3(c) is a superposition of five patterns displayed in Fig. 3.3(e), each rotated by 72° relative to each other. We bare in mind that this process is useful only because it helps identify the domain structure of the oxide layer by singling out one domain. Note further that patterns similar to Fig. 3.3(c) have been observed for annealing and oxidation temperatures between 700 K and 840 K. Such patterns have routinely been obtained for annealing times of about one hour. No further improvement in the contrast quality has been observed for longer exposure and/or annealing times. In order to confirm the oxidation of the Al atoms of i-AlPdMn only, XPS measurements were performed. In Fig. 3.5, the spectra of Al $2p$ states for the clean surface and after exposure to O_2 are displayed. Changes in the core levels of Mn and Pd could not be measured, as previously reported by Chang et al. [66]. This confirms that the new features observed in the LEED pattern after O_2 exposure of the quasicrystal arise from an Al-oxide film. Within our limited resolution, we cannot support the occurrence of Al ions in two different oxidation states or coordinations. Therefore, we were not able to determine the exact stoichiometry of the oxide film.

3.4 Structure model

As mentioned above, structure models for the growing oxide film have been proposed in the case of the oxidation of the NiAl(110) surface at high temperatures [61, 63, 70]. Basically, it consists of two (111) γ-domains in-plane oriented by ordering

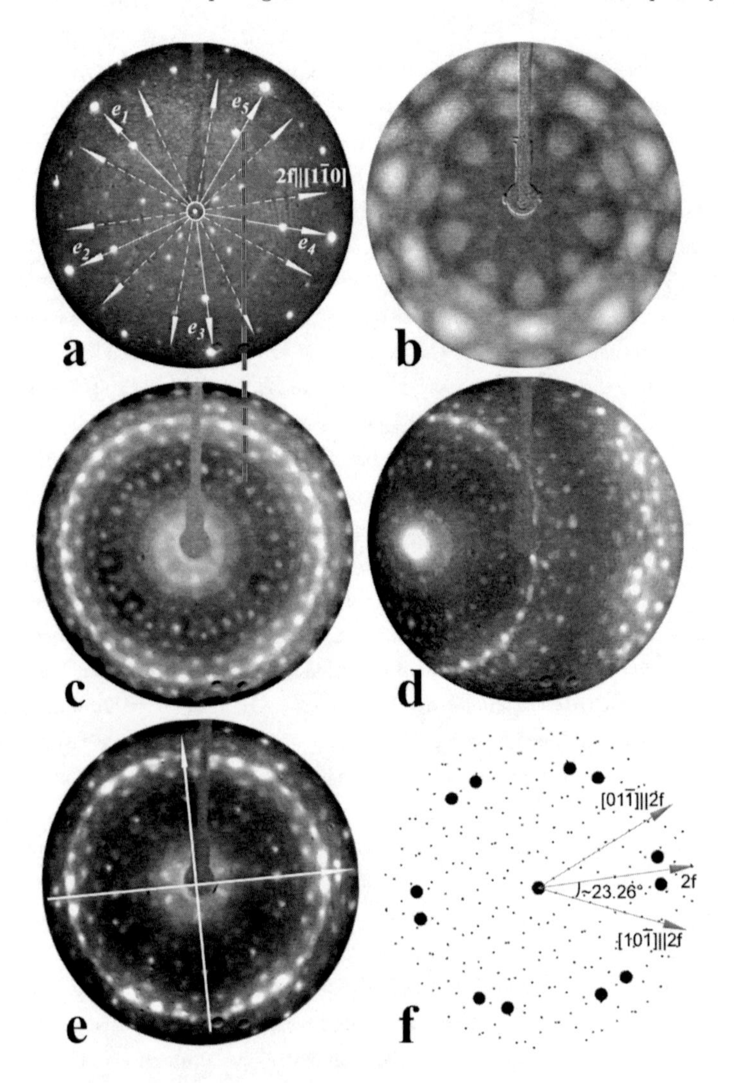

Fig. 3.3: LEED patterns obtained at a primary-electron energy of 55 eV from (a) the clean pentagonal surface of i-AlPdMn, (c) after exposure to 800 L O_2 at 700 K and subsequent annealing for 2 hours, (d) after tilting the sample by 15°, and (e) after Ar^+ sputtering the sample corresponding to (c) for 1 minute and subsequent oxidation at 700 K. The weighted reciprocal lattice of one pair of ℓ-domains is presented in (f) and oriented in accordance with the LEED patterns. SEI pattern from the clean fivefold-symmetry surface of i-AlPdMn is displayed in (b).

of the elementary oxygen cells along the $[1\bar{1}\bar{1}]$ and $[1\bar{1}1]$ directions of the NiAl substrate. As displayed in Fig. 1 of [70], the two alumina domains have each an azimuthal separation angle of 24° with respect to the $[1\bar{1}0]$ direction of the intermetallic alloy and a total separation angle of 48° with respect to each other. The in-plane separation angle between the oxide domains is the same in all proposed structure models. Differences appear in the exact atomic positions inside these cells. Since the oxide layer may not necessarily have the Al_2O_3 stoichiometry and is, additionally, too thin for a unit cell be defined, and grossly strained and distorted [63], we designate each cell of this structure the ℓ phase and use it to represent the oxide layer on the quasicrystal surface. Figure 3.3(f) displays the weighted reciprocal lattices of two (111) ℓ-domains azimuthally separated by 46.5°. Weighted means that the structure factors are taken into account in the reciprocal-lattice calculation, i.e., spots which appear larger in the simulated pattern will appear brighter in the diffraction pattern [71]. The orientation and scale of the calculated pattern are chosen to match those of the diffraction experiments. All the diffraction spots arising from the alumina film present in Fig. 3.3(e) can be found in the simulated pattern.

Unlike the NiAl case, the two ℓ-domains are not separated by 48° but by 46.5° with respect to each other. This difference can be accounted for by the 0.74° angle mismatch between the $[1\bar{1}\bar{1}]$ and $[1\bar{1}1]$ directions of the CsCl structure and the in-plane twofold-symmetry axes of the i-structure, as shown in Fig. 3.4. In fact, the pentagonal surface of i-AlPdMn is closely related to (110) surfaces of CsCl-type structures with a lattice constant of about 2.9 Å, while an in-plane twofold-symmetry axis corresponds to the $[1\bar{1}0]$ direction [34]. Therefore, an orientation of the ℓ-domains determined by the in-plane twofold-symmetry axes, and hence similar to the orientation on NiAl(110), is reasonable. However, this fundamental orientation relationship is slightly modified since atomic rows in the oxide layer (in $[10\bar{1}]$ and $[01\bar{1}]$ directions, see Fig. 3.4) align with twofold-symmetry axes rather than $[1\bar{1}\bar{1}]$ and $[1\bar{1}1]$ directions of the CsCl structure. Consequently, two ℓ-domains are rotated by $48° - 2 \times 0.74° \simeq 46.5°$ with respect to each other. The bright spots present in Fig. 3.3(c) correspond to a lattice parameter of 2.59 ± 0.02 Å, which is close to the distance of 2.60 Å between two rows of O atoms in the oxide structure [60]. This argument is based on the average quasicrystalline structure deduced from its long-range properties. An atomistic interface model based on the local arrangement of atoms would therefore be not reliable for the present case and not compatible with the quasicrystalline order at the surface.

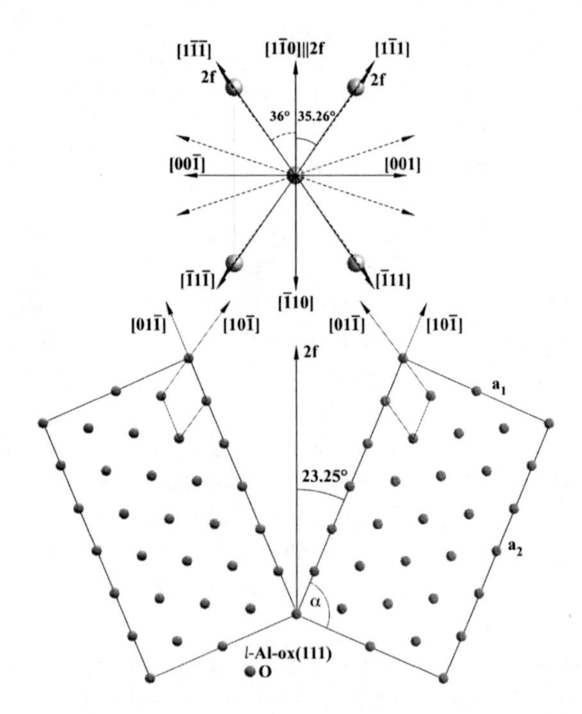

Fig. 3.4: (top) Orientational relationship between the pentagonal surface, represented by the in-plane twofold-symmetry axes (dashed arrows), and the closely related (110) surface of a CsCl-type structure (spheres, solid arrows). (bottom) Orientational relationship between a pair of ℓ-domains and the fivefold-symmetry surface of i-AlPdMn, represented by an in-plane twofold-symmetry axis. The unit cell of the oxide layer is represented by the lattice parameters a_1, a_2, and α. Elementary oxygen cells are symbolized by rhombohedra.

The reciprocal lattice of the unit cell proposed by Kresse et al. [63] was also compared with the recorded LEED pattern. Several features, as for instance the 30 bright spots, are present in the simulated diffraction pattern. However, the overall agreement is significantly lower than with the model from Lykhach et al. Therefore, we favor the model of Lykhach et al. to correspond to the oxide film present on the fivefold-symmetry surface of i-AlPdMn. The agreement of the pattern displayed in Fig. 3.3(f) with the pattern in Fig. 3.3(e) confirms that the oxide film consists of five pairs of ℓ-domains. The pairs are rotated by 72° with respect to each other. All diffraction spots observable in pattern Fig. 3.3(c) can be accounted for by this structure model. It is consistent with the fact that no preferential twofold-symmetry axis exists at the quasicrystalline surface, i.e., each of the five twofold-symmetry directions present at the fivefold-symmetry surface of i-AlPdMn should equally promote the alignment of a pair of ℓ-domains. The formation of a well-ordered alumina film on the pentagonal surface of i-AlCuFe quasicrystal was previously reported [72]. However, in this case, only one azimuthal orientation of the Al-ox domains was observed, unexpected from a quasicrystalline material in the icosahedral structure presenting seeds with multiple degrees of freedom for the alignment of a growing layer. In the light of our observations that one of the five domains can be promoted by Ar^+ bombardment, it is conceivable that the i-AlCuFe samples were mechanically manipulated in one direction prior to oxidation and/or there exists a preferential growth direction in the multi-domain quasicrystalline structure which gives rise to the growth of the oxide material in one direction.

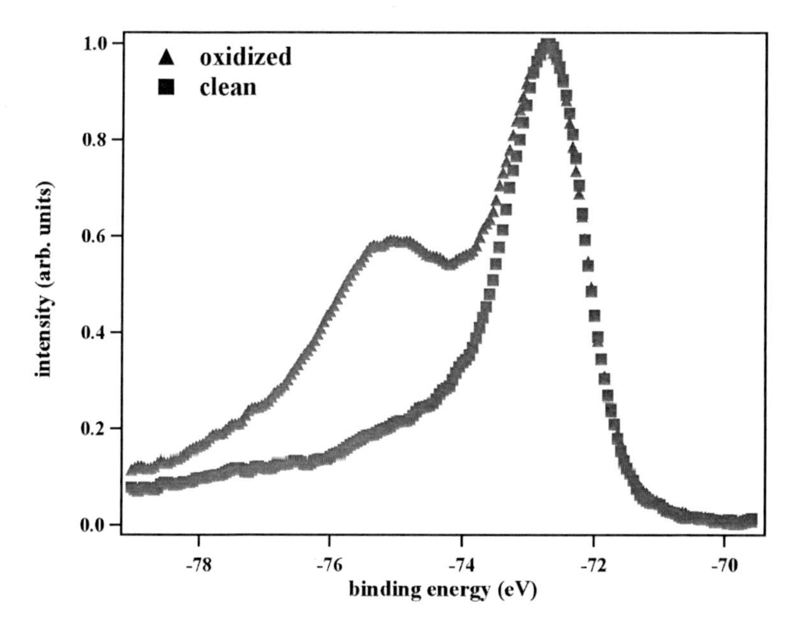

Fig. 3.5: XPS spectra of Al $2p$ core electrons from the clean surface (squares) and after its exposure to 800 L O_2 (triangles) corresponding to the LEED pattern in Fig. 3.3(a).

The electronic structure of the oxide film is still to be investigated. However, the band gap energy of the oxide films grown on $Ni_3Al(100)$ and CoAl(100) were already measured [58, 73] and found to be 4.3 and 4.5 eV, respectively. Because the structure of the Al-ox film grown on $Ni_3Al(100)$ is related to that found in this work, the band gap energy of the oxide film grown on i-AlPdMn should have a comparable value.

In summary, unlike to previous reports [64, 65], we have observed the formation of crystalline domains on the pentagonal surface of i-AlPdMn after exposure to oxygen for a substrate temperature of 700 K. Five pairs of ℓ-domains exposing their (111) faces parallel to the surface and rotated by 72° with respect to each other are formed. The orientational relationship between these domains and the substrate is a result of the close relationship between the icosahedral structure of AlPdMn and the CsCl structure. By virtue of the relative ease of preparation of the oxide film in the form of nanometric domains, we think that quasicrystals bare the potential of use as catalyst carriers of active metals without resorting to self-size-selecting processes of growing films.

Chapter 4

CdTe and PbTe nanostructures on the oxidized i-AlPdMn

As mentioned in Chapter 1, properties of semiconductor nanocrystals or quantum-dots (QDs) have received particular attention in the past decades [74]. With decreasing size, solids are gradually losing their bulk properties, approaching more and more molecular-like behavior. Due to the confinement of charge carriers to the restricted volume of the small particles, quantum mechanical phenomena are observable for which the range of sizes under investigation is frequently called the size-quantization regime. In the case of semiconductors, size-dependant effects are particularly remarkable.

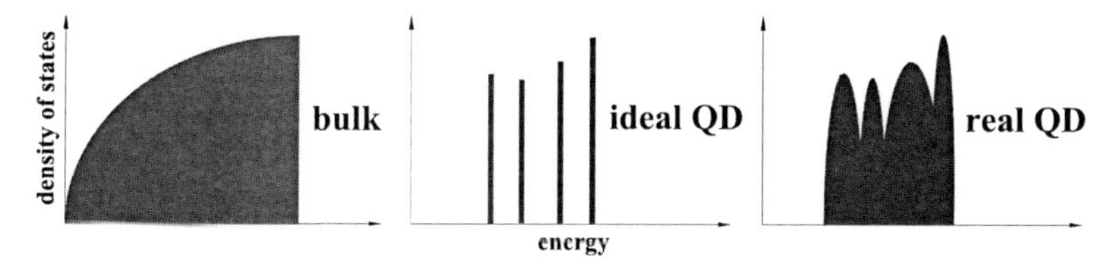

Fig. 4.1: Density of electron states in bulk and size-quantized semiconductor. The optical absorption spectrum is roughly proportional to the density of states.

When a particle (an electron in our case) is confined to a volume in space, two things happen: it acquires kinetic energy (referred to as confinement energy), and its energy spectrum become discrete [Fig. 4.1(middle)]. In a bulk semiconductor, the conduction electrons are free to move around in the solid, so their energy spectrum is almost continuous, and the electrons density of states as function of the energy increases as its square-root [Fig. 4.1(left)]. If it would be possible to produce a piece of semiconductor so small that the electrons "feel totaly" confined, their energy spectrum would become discrete and the band gap energy would increase [Fig. 4.1(right)]. Confinement effects occur in QD if their size is smaller or comparable to the exciton (electron-hole pair) Bohr radius in the material [75]. An exciton is comparable to a hydrogen atom and can be thought as the lowest excited electronic state of the bulk solid. In a QD, an electron-hole pair is also referred as exciton even if the charge carriers are bound by the confining potential rather than the Coulomb interaction.

material	exciton Bohr radius (nm)	reference
CuCl	1	[75]
Si	5	[76]
CdSe	6	[75]
CdTe	10	[77]
Ge	18	[76]
PbS	20	[75]
InAs	34	[75]
PbSe	46	[75]
PbTe	50	[75]
InSb	54	[75]

Table 4.1: Bohr radii of excitons in representative semiconductors.

As displayed in Fig. 4.1, the control of the size of the QD allow us to modify the optical properties of the semiconductor: strong absorption occurs at certain photon energies, at the expense of reduced absorption at other energies. Light always interacts with one exciton regardless of the size of the semiconductor, so that the integrated absorption does not change. Ideally, the density of states becomes a series of delta function in a QD, so those transitions must be very strong to conserve the overall absorption. The strong transitions are the result of concentration of the optical transition strength into a few narrow energy intervals. Normally, the optical properties of a material do not depend on the intensity of the light. However, at high enough intensities, materials become nonlinear, i.e., the light changes the material properties, which can then be sensed by a separate "probe" light beam or by the beam that change the material itself. With their strong absorptions, QD offer the potential for strong resonantly enhanced optical nonlinearities and are thus candidate materials for applications such as optical switching and information processing. According to Schmitt-Rink et al. [78], the linear and resonant nonlinear optical properties will exhibit the greatest enhancement when the nanocrystal radius is much smaller than the Bohr radius in the bulk material, the so-called strong-confinement limit. Despite the effort that has been performed into investigations of QDs in recent years, little experimental work has been accomplished with samples that strongly confine electrons and holes. The difficulty of reaching the strong-confinement regime becomes evident when the Bohr radii of the excitons in semiconductors are considered, which are typically around 10 nm or less [see Table 4].

To investigate the strong-confinement limit, QDs of narrow-gap materials like PbTe are desirable. The small electron and hole masses in this material imply large confinement energies, split about equally between carriers [75]. PbTe is a IV-VI, semiconductor with a narrow direct band gap of 0.32 eV at room-temperature and one of the most sensitive materials for infrared sensors. Since all technologically important narrow band gap semiconductors are difficult to fabricate and handle in bulk form, growth of thin films on a dielectric substrate, for instance an oxide, is the preferred solution for large line or area arrays. PbTe possesses a rock-salt structure with a lattice constant of 6.36 Å. CdTe, a II-VI direct semiconductor with a wide direct band gap of 1.56 eV at room-temperature, is considered as one of the most important photovoltaic materials due to its high absorption coefficient and band-gap energy, which is close to the optimum value [79]. It possesses a zincblende structure with a lattice constant of 6.48 Å. Because of the industrial significance of CdTe, investigations and productions of QDs of this material in the strong-confinement regime are highly important.

In the previous Chapter (Chapter 3), we have reported the formation of a new kind of crystal-quasicrystal interface by oxidation of the pentagonal surface of the i-AlPdMn quasicrystal. A 5 Å thick, well-ordered Al-ox film is formed, which consists of five pairs of ℓ-domains rotated by 72° with respect to each other and of a size of about 35 Å [80]. Here, we present the results obtained for the deposition of CdTe and PbTe onto the oxidized pentagonal surface of i-AlPdMn.

4.1 Crystal structures

Figure 4.2(a) displays a LEED pattern observed from the oxidized pentagonal surface of i-AlPdMn obtained after exposure to 800 L O_2 at 700 K and subsequent annealing at the same temperature. As reported in the previous Chapter, this procedure leads to the formation of a 5 Å thick, well-ordered Al-ox film, which consists of five pairs of ℓ-domains rotated by 72° with respect to each other. Each pair of domains is aligned along a twofold-symmetry direction of the pentagonal surface of i-AlPdMn. Spot-size analyses indicated an average domain size of approximately 35 Å (for details see Chapter 3).

We present here the result of the deposition of CdTe and PbTe onto the oxidized fivefold-symmetry surface of i-AlPdMn. A diffraction pattern recorded after the deposition of 40 Å of PbTe onto the alumina film at room-temperature is displayed in Fig. 4.2(b). The LEED pattern exhibits, for electrons at normal incidence, three concentric rings but no detached diffraction spots. If the sample is rotated such that the electrons incidence-angle is set to be 15° off the normal, more rings, corresponding to higher diffraction orders, appear [Fig. 4.2(c)]. Such rings were already observed, for instance, in highly-oriented pyrolytic graphite [81]. They indicate the existence of crystallites with random in-plane orientations. In case of CdTe, similar rings were observed but of slightly lower quality (not shown). For both cases, the stoichiometry at the surface was confirmed by means of AES after the deposition. Contrary to the randomness of the azimuthal orientations of the crystallites, their polar orientations are well defined, which is magnified by the high intensity and high quality of the rings.

Figure 4.3 presents a comparison between a LEED pattern obtained from the 40 Å PbTe film and a schematic representation of the reciprocal lattice of a (001) face of PbTe. The radius of the diffraction rings are in the figure displayed and are in almost perfect agreement with the distances found in the reciprocal lattice. Therefore, we argue that the PbTe crystallites have their (001) face and the CdTe crystallites (a similar study was performed in this case) their (111) face parallel to the pentagonal surface of i-AlPdMn. Spot-size analyses suggest average domains sizes of approximately 35 Å in both cases, which corresponds to the average size of the ℓ-Al-ox domains. Both materials were previously deposited on bulk sapphire substrates [79,82,83], however, in those cases domains with sizes greater than 100 Å were observed. We argue that in contrast to conventional heteroepitaxy, the size-selection process in the present system is not controlled by the lattice mismatch

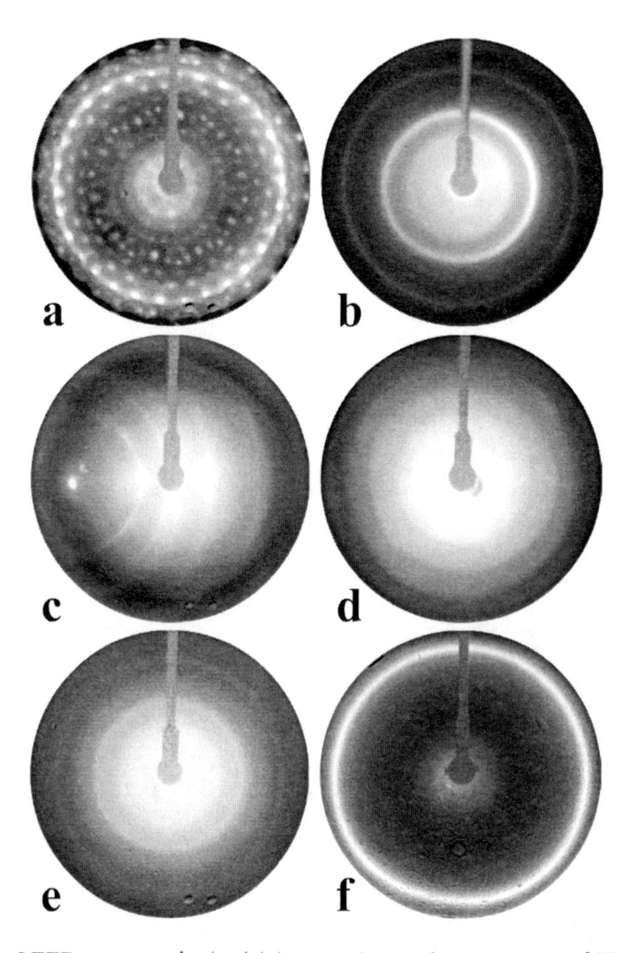

Fig. 4.2: LEED patterns obtained (a) at a primary-electron energy of 55 eV from the oxidized pentagonal surface of i-AlPdMn, (b) at primary-energy of 90 eV after deposition of 40 Å of PbTe onto the oxidized pentagonal surface of i-AlPdMn kept at room-temperature, (c) from the same PbTe as (b) but with the sample tilted 15° off the normal. An SEI pattern (d) of the 40 Å PbTe film is presented. LEED patterns (e) obtained at primary-energy of 90 eV from the PbTe film after air exposure, and (f) at a primary-energy of 64 eV after deposition of 40 Å Al onto the oxidized fivefold-symmetry surface of i-AlPdMn.

between the substrate and the growing film but by the size of the oxide domains of the substrate. Figure 4.2(d) displays the SEI pattern corresponding to the PbTe layers giving rise to the LEED pattern in (b). As in the diffraction pattern, several rings are observed, confirming the randomness of the azimuthal orientation. However, the presence of concentrical features shows that not only the surface of PbTe but also the underlying layers are ordered. The corresponding polar angles of the rings observed in the SEI patterns are compatible with the orientations described above, i.e (001) and (111) for PbTe and CdTe, respectively.

The diffraction pattern shown in Fig. 4.2(e) was, as (b), recorded after deposition of 40 Å PbTe onto the oxidized pentagonal surface and subsequent exposure to air by removing the sample from the UHV chamber. Afterwards, the sample was remounted inside the vacuum chamber in order to perform LEED. The same diffraction rings as in (b) are observed, however, with a slightly lower intensity. The fact that LEED pattern can be observed after air exposure indicates that the surface is very stable, thus resembling behavior of noble metals such as Au [84].

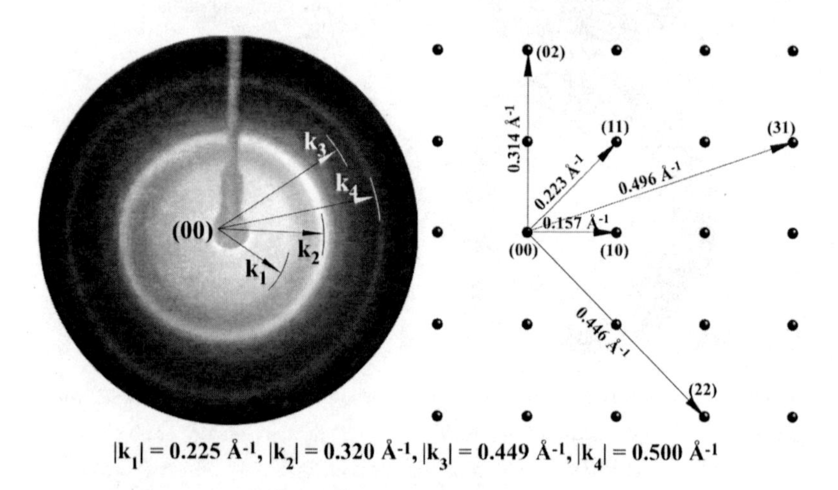

Fig. 4.3: (left) LEED pattern obtained at primary-energy of 90 eV from a 40 Å PbTe film. Four characteristic reciprocal space vectors and there respective length are also displayed. (right) Schematic representation of the reciprocal lattice of a (001) face of PbTe assuming a lattice constant of 6.36 Å.

4.2 ARPES measurements

As mentioned above, confinement effects occur in semiconductor QDs if their size is smaller or comparable to the exciton Bohr radius in the material [75]. For CdTe and PbTe, the bulk exciton Bohr radii were calculated to be approximately 100 and 500 Å, respectively [75,77], implying that, if the domains observed on the oxide film are isolated, strong confinement of the charge carriers occurs in all three dimensions. In case of wetting of the substrate by the deposited material, quantum-well (QW) states should be observable. Because the size of the nanoparticles normal to the substrate surface can be controlled by choosing the quantity of the evaporated material without any observable structural modification between one and several monolayers (ML), it would be possible to tune the size effects. The air stability of the PbTe films deposited on the oxidized pentagonal surface of i-AlPdMn permitted us to transfer the sample to an other UHV chamber equipped with an ARPES apparatus. This allows us to investigate the electronic density of states in a region near the Fermi level (E_F), in order to probe features arising from size effects.

Figure 4.4 displays photoemission spectra of the valence band of a 40 Å PbTe films corresponding to the diffraction pattern Fig. 4.2 (e). The three spectra were taken at three different polar angles Θ 0, 20, and 40° off the normal. The photons incidence azimuthal angle was not set along a specific direction in this experiment due to the randomness in the azimuthal orientations of the PbTe domains as described above. The density of states at E_F and up to 2 eV is almost reduced to zero. Compared to the bulk band gap of PbTe which is 0.25 eV at room-temperature, the band gap here is much larger, at least 2 eV. This is probably due to the isolation properties of the underlying Al-ox film, which acts as a "battery" and provoke a virtual "shift" of E_F.

As emphasized by the solid vertical lines in the figure, the maxima do not shift in energy, implying that no dispersion of the different peaks could be observed. We argue that either the film was corrupted by the exposure to air, which was already reported for PbTe films [85,86], or confinement effects in all three dimensions prevents an observations of an angular dependance [15]. Because, the surface free energy of PbTe [87] is smaller than of Al_2O_3 for instance [88], we anticipate that the PbTe film grown on the oxidized fivefold-symmetry surface of i-AlPdMn should be two-dimensional [11]. In this case, confinement in only one dimension should be observe and in the ARPES investigations an angular dependence measured. Therefore, we conjecture that the electronic properties of the PbTe film was influenced by atmospheric contamination, i.e., O_2, CO, and CO_2.

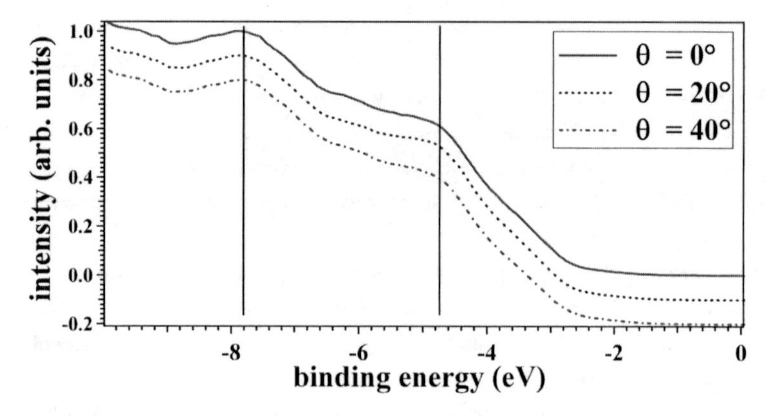

Fig. 4.4: ARPES spectra obtained, from a 40 Å PbTe film grown on the oxidized pentagonal surface of i-AlPdMn, at three different emission angles. Vertical lines are here to emphasize, that the maxima do not shift in energy, implying the absence of dispersion.

4.3 Ag on i-AlPdMn and d-AlCoNi

In order to probe the relationship between atomic ordering, dimensionality and electronic properties in crystal-quasicrystal hybrid structures, we have investigated by angle-resolved photoemission the *sp*-valence levels of Ag films grown on d-AlCoNi and i-AlPdMn. The experiments were performed at the VUV-beamline of the synchrotron source Elettra in Trieste, Italy.

Fig. 4.5 reports photoemission spectra of the clean fivefold-symmetry and tenfold-symmetry surfaces of i-AlPdMn and d-AlCoNi, respectively, within the first 10 eV below E_F at near normal emission, to compare their electronic structure. In both cases, the spectra are dominated by the emission from the transition metal *d* levels. For d-AlCoNi the feature centered at 1.8 eV corresponds to the superposition of Co and Ni 3*d* states, while the weak emission at nearly 6 eV is mainly due to Al *sp* states. For i-AlPdMn the two structures at 0.8 eV and at 4.5 eV below E_F originate from Mn 3*d* and Pd 4*d* states, respectively. Here, the assignment is accomplished by exploiting the properties of the photoemission cross section [89] as a function of the photon energy.

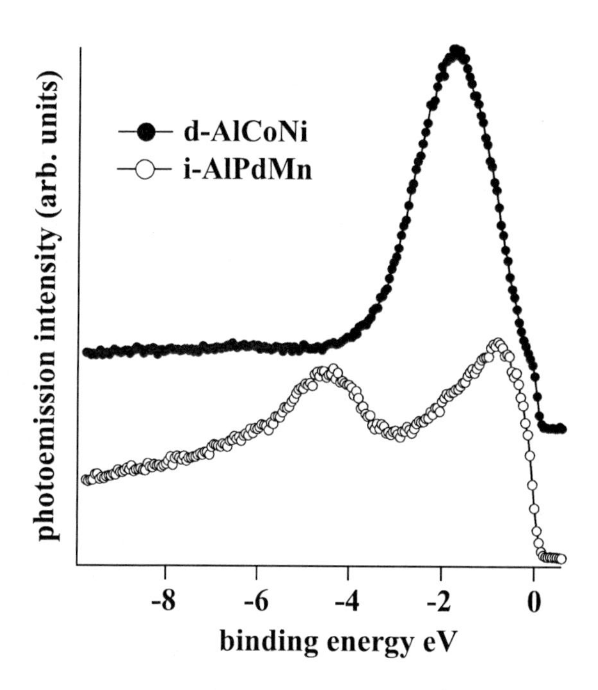

Fig. 4.5: Energy distribution curves for d-AlCoNi (upper spectrum, $h\nu = 130$ eV) and i-AlPdMn (lower spectrum, $h\nu = 156$ eV).

Ag was evaporated from a power regulated atomic source onto the quasicrystalline surfaces kept at 140 K. Successive gradual annealing to room-temperature was performed to facilitate a layer-by-layer growth [47]. This procedure critically affects the film morphology by improving the film-height uniformity, thus making quantum size effects readily detectable by photoemission. The result for 7 ML Ag (1 ML= 2.36 Å) on d-AlCoNi is shown in Fig. 4.6(d). In the upper spectrum (low-temperature deposition) two QW states emerge, as a consequence of the electron confinement along the surface normal. These peaks can be assigned to the integer quantum number n, for which $k_\perp$, the wave-vector component perpendicular to the surface, fulfils the relation 4.1, where Φ_S and Φ_I are the phase shifts at surface and interface, respectively, and t the film thickness.

$$2k_\perp t + \Phi_S + \Phi_I = 2\pi n, \tag{4.1}$$

If t is atomically uniform, Eq. 4.1 gives a single set of wave vectors and corresponding energies, thus resulting in well defined peaks in the photoemission signal. On the other hand, when several film thicknesses are simultaneously present over the experimentally probed area, as observed for room-temperature deposition, a large number of electronic states will be observable according to Eq. 4.1. In this case, the energy distribution of the QW states is broadened, as displayed in the lower spectrum of Fig. 4.6(d). The $n = 1$ and $n = 2$ peaks approximately fall at the same energy as for the QW states of the upper spectrum, since the island-height distribution is centered approximately on the nominal coverage, yet with diminishing spectral weight. The two spectra differ also, significantly in the region around E_F, where only for low-temperature growth the Ag(111)-derived surface state is observed. This state, which is created at the surface in a gap of the bulk band structure formed by the termination of the crystal at the surface, is called a Shockley state. Its formation can be modeled by adding a weak potential to the free electron gas of a solid, what is generally applied for the description of sp-bands in metals.

The morphologies of Ag layers grown on d-AlCoNi and i-AlPdMn have been previously studied for a wide thickness range [29,90]. STM images show that room-temperature deposition gives rise to rotationally organized fcc islands exposing their (111) faces parallel to the substrates surface. Thirty equally spaced LEED spots lying on a ring were interpreted as originating form five different in-plane orientations of nanocrystalline domains. Here, diffraction patterns obtained for low-temperature deposition of 7 ML Ag on d-AlCoNi [Fig. 4.6(a-c)] display circles centered around the surface normal, with decreasing radius as the primary electron energy is increased. The formation of featureless rings instead of spots points out the presence of azimuthal disorder in the Ag film (as for PbTe and CdTe on the Al-ox film, see above). The separation between first and second order diffraction rings [Fig. 4.6(c)] is compatible with the in-plane interatomic distance of Ag(111) planes (2.89 Å), while from the width of the rings an average domain size of 50 ± 10 Å is evaluated. Similar diffraction patterns, but with inferior intensity contrast, were obtained for room-temperature deposition of the same amount of Ag (not shown).

The high spectral definition of QW states achieved in case of low-temperature growth allows us to investigate the dependence of the electronic structure in crystal-quasicrystal interfaces as function of $k_\parallel$, the wave vector component parallel to the surface. Photoemission spectra for 7 ML Ag film grown on d-AlCoNi are displayed for $k_\parallel$ along the [$0 0 1 \bar{1} 0$] [Fig. 4.7(a)] and [10000] [Fig. 4.7(b)] axes of the substrate,

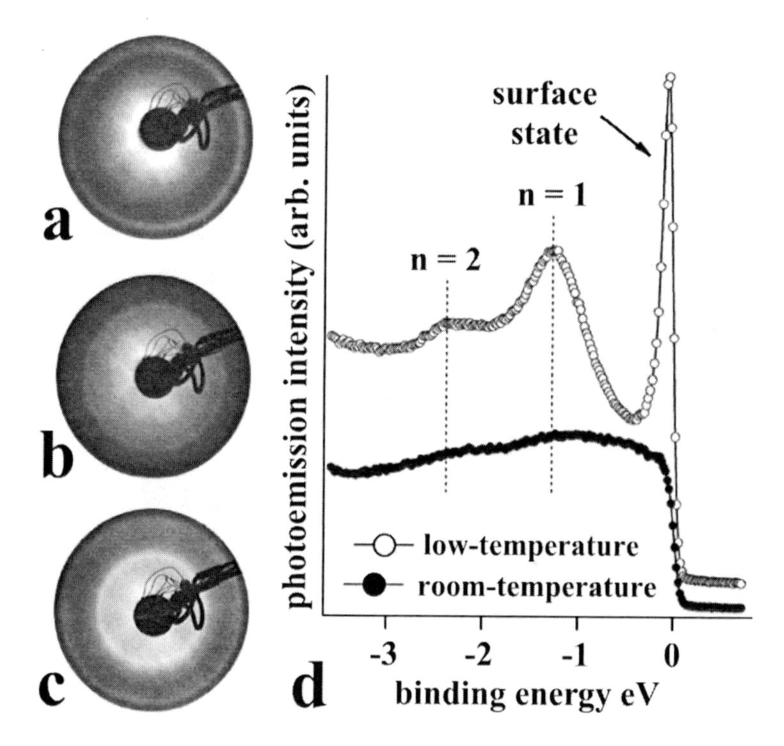

Fig. 4.6: LEED patterns obtained for 7 ML Ag on d-AlCoNi at (a) 52, (b) 79, and (c) 139 eV. (d) Energy distribution curves at, $h\nu = 51$ eV for the same coverage deposited at 140 K (upper spectrum) and room-temperature (lower spectrum).

i.e., two inequivalent twofold-symmetry axes of the decagonal surface. Along both in-plane directions the *sp*-valence levels exhibit a highly dispersive character, in spite of the rotational disorder of the metal layer which generally points towards localization. This suggest that the presence of domain boundaries in the film morphology has negligible influence on the electronic states, since the local crystalline arrangement dominates over the microscopic island distribution. Similar observations have recently been made on the valence band of graphite, where band dispersion coexists with azimuthal disorder [91]. While $n = 1$ and $n = 2$ states can be followed up to E_F, the $n = 3$ state is detectable only for off-normal geometry. This observation is a consequence of lifetime broadening, which increases as function of the binding energy and also explains the difference in the spectral shape of $n = 1$ and $n = 2$ states.

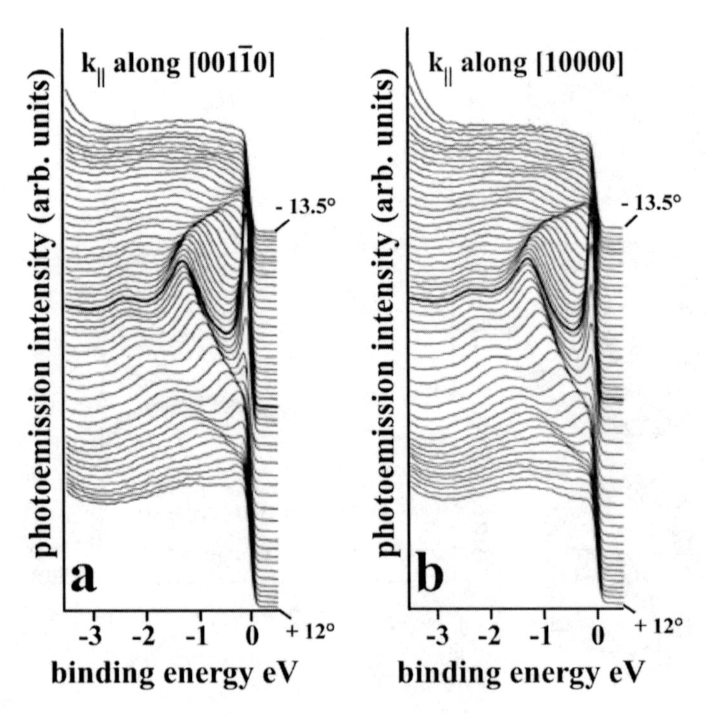

Fig. 4.7: Energy distribution curves for 7 ML Ag film grown on d-AlCoNi, as function of $k_{\|}$ scanned along two major quasicrystalline directions (see text for more details).

We have used the spectra shown in Figs. 4.7(a) and (b) to generate the two-dimensional plots displayed in Figs. 4.8(a) and (b), respectively, where the gray scale is proportional to the intensity of the photoemission signal. The band dispersion can be quantitatively described in terms of the effective mass m_n^*, as a multiple of the electron mass m_e. Black lines are the best parabolic fits to the experimental points (open circles) in a region of ± 0.2 Å^{-1} near normal emission ($k_{\|} = 0$). For each state the effective mass is basically the same along the two axes within the experimental error ($\pm 0.05\ m_e$, mainly originating from the uncertainty in $k_{\|}$), as one would expect for Ag(111) *sp*-derived bands [92]. Furthermore, rotational disorder contributes to average the band dispersion over the azimuthal angle. The almost perfect free-electron character of the bands clearly shows up if the parabolic fits are extended to larger $k_{\|}$ values. In terms of interaction between film and substrate states this behavior indicates that the interface potential barrier reflects all Bloch-type *sp*-states of the Ag layer.

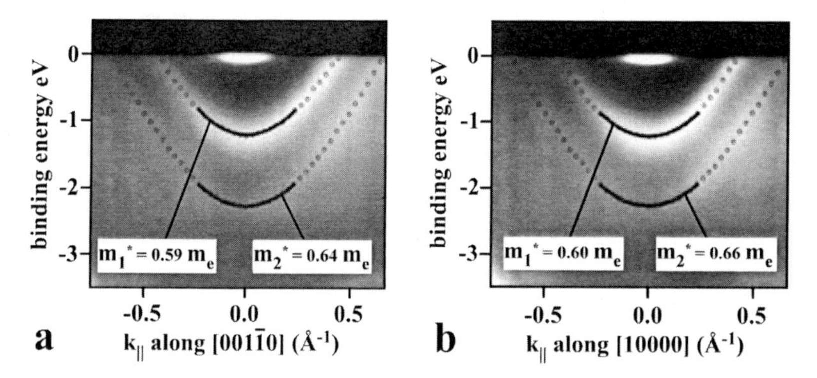

Fig. 4.8: Intensity plots derived from the two sets of spectra reported in Fig. 4.7.

QW states exhibit similar properties for Ag films grown on i-AlPdMn at low temperature. The upper part of Figs. 4.9(a) and (b) show LEED patterns for the clean pentagonal surface and 7 ML thick Ag film, respectively. The spot distribution and in-plane relations between symmetry axes match perfectly with previous observations for room-temperature growth [29]. Fig. 4.9(c) is an intensity plot generated from the photoemission spectra recorded for the 7 ML Ag film along the [001$\bar{1}\bar{1}$1] substrate axis. Again, the parabolic dispersion of the QW state indicates a featureless interface potential extending over several eV below E_F, which acts as an highly reflecting barrier for Ag electrons. Figure 4.9(d) displays an intensity plot for a 14 ML thick Ag film grown on i-AlPdMn. We observe that the parabolic dispersion is retained and the binding energy of the QW states decreased, in agreement with Eq. 4.1, while the changes in effective mass can be attributed to the dependence of $k_{\perp}$ on $k_{\|}$, as shown for Cu films on Co (001) [93].

On the basis of the photoemission results for Ag layers grown on the two quasicrystalline substrates we can address the origin of quantum size effects in these and related systems. We first note that the QW states in the two systems under investigation display very strong similarities, in contrast to the markedly different electronic structure of the quasicrystalline substrates (Fig. 4.5). The observation of QW states with the same parabolic band dispersion in the two systems indicates that the electron confinement is largely independent on the density of states of the quasiperiodic surface, acting as a template for film growth.

Electron quantum confinement of the valence levels was identified as the driving force for the height selectivity of Ag and Bi islands on icosahedral substrates [94]. Electronic growth is associated to the oscillatory thickness dependence of the density of states at E_F, that gives rise to a corresponding behavior in the electronic term of the total energy [95]. This results in the preferential formation of magic height islands. The formation of QW states derived from Ag and Bi *sp*-levels was attributed to the suppression of *sp*-density of states near the E_F in the substrate, and therefore considered as a direct manifestation of the pseudogap. The present results give indeed a direct prove that QW states develop in this class of systems, thus supporting electronic growth as the basic mechanism for the observed height selectivity of the islands. However, they demonstrate that the presence of a pseudogap in the total or partial *sp* density of states (the latter is expected to exhibit the same pseudogap of the former due to hybridization between *sp* and *d* levels) is not required for attaining electron confinement in these systems. The photoemission data show that QW states form also well outside the region of the substrate pseudogap, which typically extends over just a few tenths of an eV around E_F [96]. QW states are found distributed over a wide binding energy range and appear to be insensitive to the specific features of the density of states of the quasicrystalline substrates.

We argue that the strong electronic confinement in these systems is mainly driven by the symmetry of the wave functions, which governs the coupling between the electronic states of the film and the substrate. The role of the electronic state symmetry has been already established for films grown on crystalline substrates in a number of cases [97, 98]. Bloch waves in the Ag layer obey point-group symmetries derived from the crystalline atomic arrangement. It is still controversial whether wave functions in quasicrystalline alloys are fully localized, critical, or extended. In all cases, they respond to crystal-forbidden symmetries. Electronic coupling, then, is strictly forbidden due to the lack of common symmetry of the wave functions across the interface for every electron energy and wave vector. Therefore, the formation of magic height islands, rather than being associate to the presence of a pseudogap near E_F, appear to be a general consequence of the incompatible symmetries of electronic states in crystalline and quasicrystalline materials.

In this Chapter, we have used the oxidized fivefold-symmetry surface as template for molecular beam epitaxy of the binary semiconductors CdTe and PbTe. In both cases, diffraction patterns consisting of several concentrical rings are observed. This

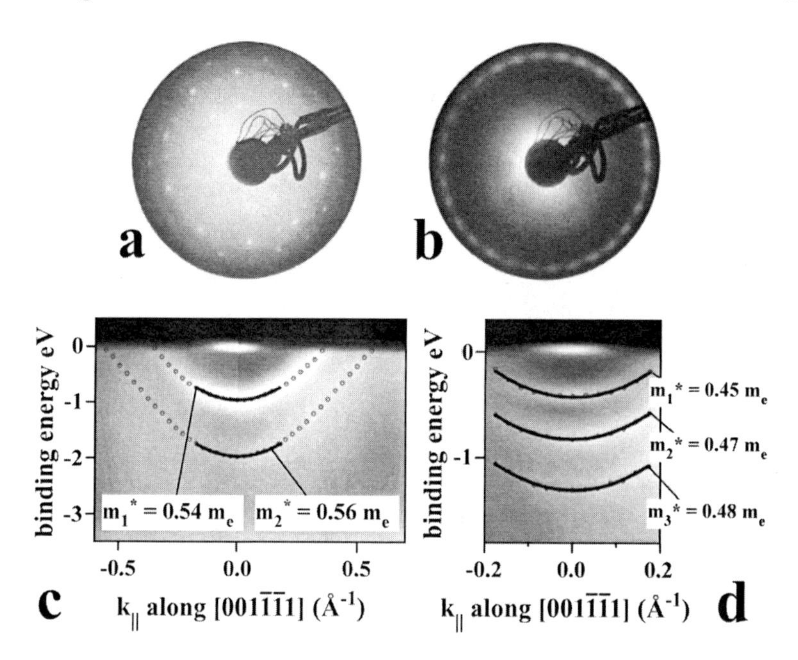

Fig. 4.9: LEED patterns obtained from (a) the clean fivefold-symmetry surface of i-AlPdMn at 74 eV and (b) from a 7 ML thick Ag film at 54 eV deposited at 140 K. The intensity plots measured along the $[00 1 \bar{1} \bar{1} 1]$ direction of the substrate for (c) 7 and (d) 14 ML Ag have different energy and $k_{\|}$ scales.

implies a randomness in the azimuthal orientations of the crystallites. The polar orientations are clearly defined; CdTe and PbTe domains expose their (111) and (001) faces parallel to the substrate surface, respectively. The Al-ox film was also used as substrate for the deposition of Al, which results in the formation of (111) Al domains of an average size of approximately 35 Å [see Fig. 4.2(f)]. We conjecture that the average domain sizes are not given as usually observed in molecular-beam epitaxy by the lattice mismatch between the substrate and the growing film, but by the size of the oxide domains of the substrate. In general, Al-ox domains formed on the pentagonal surface of i-AlPdMn have the potential to serve as templates for the production of nanoparticles with an in-plane size of approximately 35 Å. We anticipate that this technique will find wide application whenever size selection base on crystallographic constraints is not operative.

We have also reported quantum size effects arising in Ag films deposited onto the fivefold-symmetry surface of i-AlPdMn and onto the tenfold-symmetry surface of d-AlCoNi, which is an example of metal-on-quasicrystal systems. By analyzing the Ag *sp*-derived QW states, we assert that the interface with the quasiperiodic material constitutes an efficient barrier for electron propagation, due to lack of common point-group symmetries between Bloch-like and critical wave functions.

Chapter 5

Si deposition onto i-AlPdMn: effects on the surface Debye temperature

Since preparation methods of flat and well-defined quasicrystalline surfaces have been established and can be readily applied, quasicrystals became potential substrates for heteroepitaxy, opening new promising perspectives. In atomic-beam epitaxy, the lattice mismatch between the substrate and the adsorbate dictates the growth mode of the surface film. A small mismatch leads to pseudomorphic growth, while a larger mismatch results in a domain structure of the growing film, which often consists of self-size-selecting islands possibly leading to the formation of quantum dots. During the last few years, several groups have been investigating thin film growth on quasicrystals. For instance, the growth of Al nanostructures on the fivefold- and threefold-symmetry surfaces of i-AlPdMn has been studied by Lüscher et al. [99]. They observed five different domains of Al fcc nanocrystals in rotational alignment with the substrate symmetry. Recently, Weisskopf et al. have grown Fe, Ni, and Co in cubic domains on the pentagonal surface of i-AlPdMn [34, 100, 101] and demonstrated the magnetic character of these domains. Therefore, Si, Ge and other semiconductor materials deposition on the fivefold-symmetry surface of i-AlPdMn was a logical step in the study of heteroepitaxy on quasicrystals. Ideally, Si would break in small crystalline domains in the nm range, as the ferromagnetic materials (see above), with the hope that some size effects will then occur.

5.1 Si growth mode on i-AlPdMn

During the continuous deposition of Si onto the fivefold-symmetry surface of i-AlPdMn kept at room temperature, the intensity of the diffraction spots arising from the quasicrystalline order decreases until they completely vanish for deposition of more than 1 ML, indicating the evolution of an unordered Si thin film. Contrary to the deposition of Al on the pentagonal surface of AlPdMn, where new LEED spots appear after the deposition of more than $4 - 5$ Å, no new LEED spots are observed after Si deposition exceeding 1 ML. Additionally, SEI investigations on Si film up to a thickness of 100 Å were performed and no ordered structure was found. These observations confirm an unordered growth of Si film on AlPdMn. This unordered growth regime was observed for a broad range of substrate temperatures ($200 - 300$ K). The same behavior was encountered in case of deposition of Ge onto the same substrate. The amorphous growth mechanism was confirmed by STM measurements performed by Ledieu et al. [102]. They observed for coverage lower than 0.25 ML a pseudomorphic growth and for larger coverage an unordered growth. They found similar results for the growth of Si on the tenfold-symmetry surface of d-AlCoNi [103], as we did.

The structures of amorphous (a-)Si and a-Ge are characterized by the absence of long-range order. Because the volume irradiated exceeds by far the volume in which some order reigns, diffraction experiments provide incoherent patterns. Analysis of the Fourier transformation of the angular intensity distribution leads to spherical symmetry of the short-range order [104]. Richter and Breitling [105] found a coordination number equal to 4, like in crystal, for the first neighbors in a-Si and a-Ge and therefore a short-range tetrahedral symmetry is most probable. They propose that a-Si and a-Ge consist of chains of tetrahedra rotated statistically about their common bonds, packed together rather regularly in domains with layer structure and separated by completely disordered regions. Electronic structure investigations on a-Si and a-Ge have shown that in comparison to their crystalline phase, the energy gap is slightly increased. The most important differences are to be found in transport properties. They are p-type, low-mobility semiconductors with a changes in conduction mechanism with temperature. However, at high temperature their transport properties approach those of intrinsic crystalline Si and Ge. In heteroepitaxy like Si/metal, for instance, a disordered growth mode for Si is the most common case if the substrate is kept at temperature lower than roughly 600 K [106]. It appears that for this temperature range the mobility of the atoms is not sufficient to overcome the high directionality of the bonds and therefore prevent the formation of long-range order. At higher temperature, crystalline Si is observed [106]. Unfortunately in our case, above 370 K, AES and XPS investigations showed that Si noticeably diffuses into the substrate. At these deposition conditions, LEED patterns showed fivefold-symmetry similar to that obtained from the clean surface of AlPdMn indicating that diffusion of Si does not disturb the quasicrystalline order at the surface (Fig. 5.1). A similar LEED pattern was obtained by deposition of 20 Å Si at room temperature and subsequent annealing at 380 K. The observation of diffusion of Si at temperature lower 600 K prevent us to achieve the required condition mentioned above to achieve growth of crystalline Si.

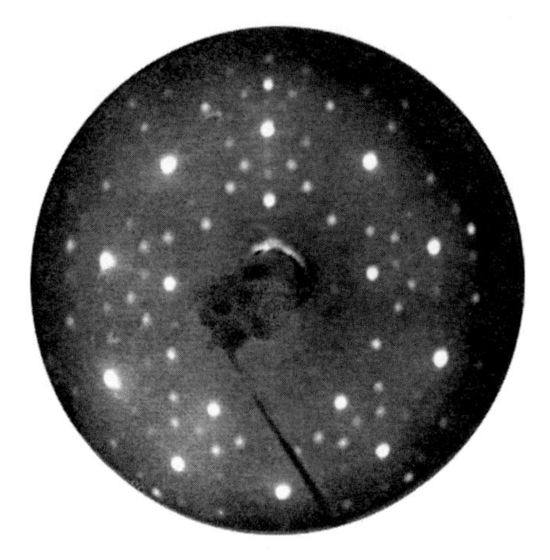

Fig. 5.1: LEED pattern obtained from the fivefold-symmetry surface of i-AlPdMn at 76 eV primary-electron energy after the deposition of 30-Å Si at 620 K.

For deposition onto the quasicrystalline substrate kept at room temperature, intensities observed in the LEED pattern arising from the pentagonal surface of i-AlPdMn decrease with the Si coverage. In order to investigate the interaction between the Si atoms and the substrate, XPS measurements were performed. Figure 5.2 displays spectra of the electronic density of states near E_F taken from the clean fivefold-symmetry surface of i-AlPdMn, after deposition of 15 Å Si, after annealing this film at 400 K and let diffuse Si into the quasicrystalline substrate, and from a Si single-crystal sample placed in the same UHV chamber. The spectrum from the clean quasicrystalline surface, possess features like a low density of states just below E_F, a small peak at approximately 2 eV arise from Mn, and an intense peak at approximately 5 eV from Pd. They are all characteristic for i-AlPdMn and were previously reported [107]. The deposition of 15 Å Si provoke an increase in the density of states around 2 eV, no further differences could be observed in this energy region. After annealing the film at 400 K for 5 minutes, these features disappear. The spectra before Si deposition and after the annealing process are then almost not distinguishable. With their STM investigations Ledicu et al. could identify preferential adsorption site for Si which should correspond to the position of Mn atoms at the surface [102]. Such assumption could be compatible with the increase of the density of states at approximately 2 eV. However, this features can also be explained simply by the presence of Si at the surface. In Fig. 5.2 a spectrum taken from a Si crystalline sample is also displayed. The features observed in this spectrum at approximately 2 eV can explain the increase in the density of states at this energy in case of a-Si on i-AlPdMn. Furthermore, in case of surface alloying between the Si and Mn atoms, a change in the core levels of Mn should be observed. We did not find any core level shift for Mn, Pd or Al. We argue, therefore, that Si does not alloy with any of the constituents of i-AlPdMn and that the increase in the density of states at approximately 2 eV below E_F, observed by means of XPS, is only due to the presence of a-Si at the surface.

In summary, we have observed that deposition of Si onto the pentagonal surface of i-AlPdMn kept below 370 K follows an unordered growth regime and leads to the formation of an a-Si film. Subsequent annealing or deposition at higher temperatures results in the diffusion of Si into the bulk of i-AlPdMn without disturbing the quasicrystalline phase at the surface.

5.2 Debye temperature measurement

5.2.1 Theoretical background

In thermodynamics and solid state physics, the Debye model is a method developed by Peter Debye in 1912 to estimate the phonon contribution to the specific heat in a solid. It treats the vibrations of the atomic lattice as phonons in a box, in contrast to Einstein model, which treats the solid as many individual, non-interacting quantum harmonic oscillators.

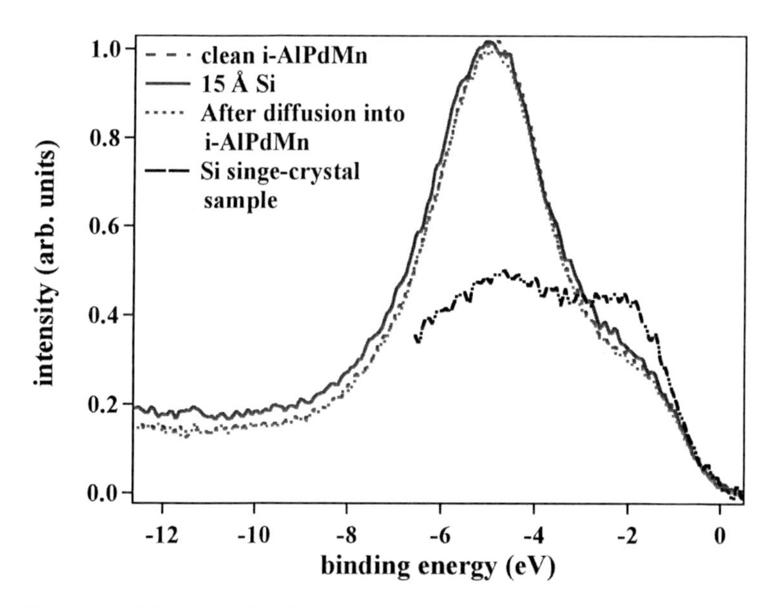

Fig. 5.2: XPS valence band spectra from the clean fivefold-symmetry surface of i-AlPdMn, after deposition of 15-Å Si, after diffusion of Si, and from a Si single-crystal sample

In crystals, phonons possess a well-defined frequency w and wave vector $\overrightarrow{k}$. In the Debye approximation, the dispersion relation can be written as $w = v\mathbf{k}$ which implies a constant speed of sound in the solid. If in the crystal N unit cell are present, there are N acoustic phonon modes. It leads to the definition of a cut-off frequency w_D and a cut-off wave vector k_{D}. The total lattice energy can then be written as:

$$U = 9Nk_{\mathrm{B}}T\left(\frac{T}{\theta_{\mathrm{D}}}\right)^3 \int_{x_{\mathrm{D}}}^{0} dx \frac{x^3}{e^x - 1}, \tag{5.1}$$

where $x_D = \hbar w_{\mathrm{D}}/k_{\mathrm{B}}T = \theta_{\mathrm{D}}/T$ with k_{B} the Boltzmann constant. This equation defines the Debye temperature θ_{D} [108].

The Debye temperature is an important microscopic parameter related to the stability of the material. It is basically a measure of the vibrational response of the material and, therefore, intimately connected with physical properties based on the lattice dynamics. θ_{D} is, for instance, used to interpolate the temperature dependance of different crystal characteristics related to atomic and ionic motion like the Debye-Waller factor, the heat capacity, the thermal conductivity, etc. Different methods can readily be applied to determine θ_D of the bulk material. Here we take advantage of the surface sensitivity of low-energy electrons to determine the surface θ_D by means of the temperature dependance of the diffraction intensities recorded in LEED experiments.

Intensity profile studies of reciprocal-space projection pose a complex problem already for ordinary crystals and even more complex for quasicrystals due to the infinite dimension of their unit cell and, consequently, the great amount of diffraction spots obtained from a well-prepared quasicrystalline surface. Beside the symmetry of the surface or the bulk, several different information, such as the domain size, the flatness of the surface, or the overall quality of the structure, can be gathered from results of diffraction experiments. Furthermore, the fluctuation of atoms around their equilibrium position at a finite temperature implies an exponential decrease in the intensity of the reflected particles (here electrons) with increasing temperature. This can be expressed as: $I \propto e^{-2M}$, where M is the Debye-Waller factor. In the high-temperature limit (typically higher than 100 K), M and the atomic mean-square thermal displacement can be formulated as:

$$2M = \frac{16\pi^2 \langle u^2 \rangle cos^2\phi}{\lambda^2} \quad \text{and} \quad \langle u^2 \rangle = \frac{3\hbar^2 T}{m k_\mathrm{B} \theta_\mathrm{D}^2}, \tag{5.2}$$

where λ is the wavelength of the diffracted particles in the material, ϕ the angle of incidence relative to the surface normal, and m the atomic mass of the scattering centers. The thermal displacement of surface atoms can be different from that of bulk atoms, what may imply a lower θ_D of the surface than of the bulk. In order to measure the surface θ_D, electrons with a short mean free path are selected as primary particles. In LEED, the primary-electron energy is kept in the range of $30 - 250$ eV, thus leading to a surface sensitive experiment. Lüscher et al. [109] have used the same procedure to obtain θ_D of the fivefold-symmetry surface of i-AlPdMn and have found 298 ± 7 K, while Colella et al. [110] have, by means of an x-ray diffraction experiment, determined θ_D for the bulk of the i-AlPdMn and found 312 K. Lüscher et al. [111] have also shown that the diffusion of Al into i-AlPdMn induces an increase of the surface θ_D by more than 100 K for an amount of absorbed Al corresponding to a 200-Å thick layer. In the following, the results obtained for the deposition and absorption of Si films onto the pentagonal surface of i-AlPdMn are reported.

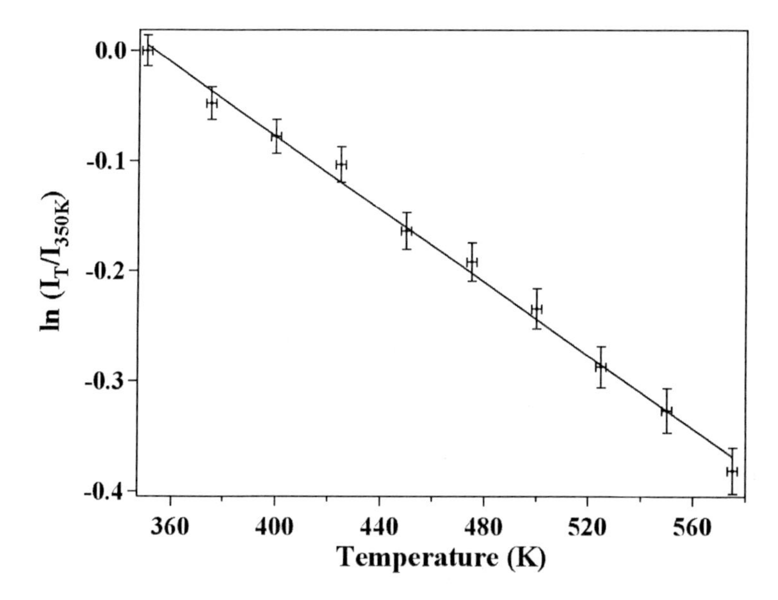

Fig. 5.3: Plot of the logarithm of the relative intensity of the specular diffraction at 43 eV primary-electron energy and 7° angle of incidence as a function of temperature, obtained after 60-Å Si absorption.

5.2.2 Experimental details

As described before, a three-grid back-view display system operated with a beam current in the low μA range was used for the LEED experiment. The diffraction intensities were collected on a home-made fluorescent screen and recorded with a 16-bit CCD camera. The temperature-dependant response of the CCD sensor was removed before the analysis of the recorded intensities. The angle of electron incidence was set at $7.1 \pm 0.2°$ relative to the surface normal, i.e., the fivefold-symmetry direction. The azimuthal alignment of the specimen was such that a twofold-symmetry plane of the bulk coincides with the scattering plane. For this scattering geometry, the primary-electron energy for the maximum intensity of the specular beam was found to be 43 eV. As in ordinary crystals, the distance between atomic planes in quasicrystals widens with increasing temperature [109]. Consequently, conditions for constructive interference change and the diffraction peaks "move" to lower energies. In order to compensate this shift, the primary-electron energy is reduced by 0.1 eV/27 K in the temperature range of $350 - 600$ K [111]. To account for the electron acceleration inside the material, an inner potential of 12 eV was chosen, assuming 4.5 eV for the work function and taking 7.5 eV for the width of the valence band of the alloy obtained from photoemission experiments [89], resulting in $\phi = 6.24°$ and $\lambda = 1.66$ Å inside the material. An average atomic mass of the sample of 7.575×10^{-26} kg was previously suggested [112] and is used in Eq. 5.2.

5.2.3 The surface Debye temperature as function of the incorporated amount of Si

The LEED pattern presented in Fig. 5.1 was recorded after the deposition of 30-Å Si onto a freshly prepared fivefold-symmetry surface of i-AlPdMn at a substrate temperature of 620 K in order to facilitate diffusion. Similar to Al [111], we observe that the diffusion of Si into the bulk does not affect the quasicrystalline order at the surface.

Figure 5.3 shows the logarithmic plot of the relative intensity of the (00) beam as a function of temperature, obtained after the diffusion of 60-Å Si into i-AlPdMn. The plot is linear in the temperature range of $300 - 550$ K. From the slope of the linear fit, we obtain a θ_D of 330 ± 7 K. The measurement was repeated for different amounts of Si and the resulting values of θ_D are presented in Fig. 5.4. Each data point was obtained after the exposure of a freshly prepared surface to Si. The θ_D increases linearly for increasing amounts of absorbed Si up to 30 Å and seems to grow slower for amounts greater than 40 Å, possibly reaching a saturation for even larger amounts of incorporated Si. The deposition of more than 60 Å was not attempted because prolonged deposition times (the deposition time is already one hour for 60 Å) would lead to contamination of the reactive Si films in the vacuum chamber. We find that, independent of how much Si was previously deposited, the freshly prepared surfaces have all the same initial value of θ_D, which was found to be 303 ± 7 K. This value for θ_D of the pentagonal surface of i-AlPdMn is comparable with that previously obtained [109].

The first step in the preparation procedure of a clean quasicrystalline surface is, as mentioned in Chapter 2, ion sputtering, which modifies the chemical composition of the surface layers due to preferential depletion of Al and favors the formation of cubic AlPd domains. The annealing process restores the nominal composition of the quasicrystal by virtue of diffusion of Al from the bulk to the surface, which

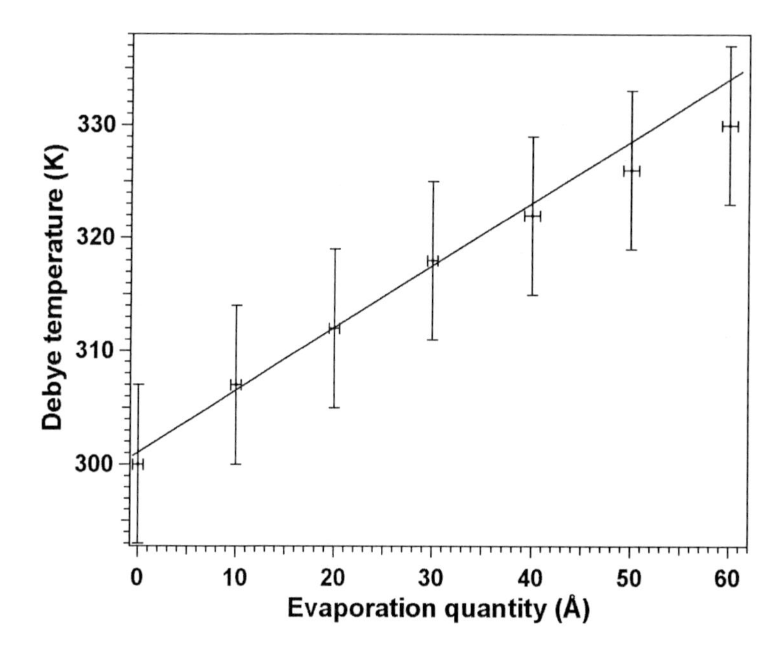

Fig. 5.4: The surface θ_D as function of the absorbed amount of Si. The solid line is a linear fit of the data points for 0 – 30 Å, suggesting the linear increase of the surface θ_D in this Si absorption range.

changes, however, the bulk Al concentration only in the range of some parts per million per cycle [111]. The process involved in the structure transformation at the surface during the thermal annealing has been studied and described [113] in terms of diffusion of vacancies toward the surface as well as diffusion of alloy constituents. Combining these two effects and considering the fact that θ_D rises with the amount of absorbed Si, while diffusion of Si into the substrate does not affect the quasicrystalline order, we conjecture that Si diffusion into i-AlPdMn is preferentially a vacancy-mediated process.

The vacancy-mediated diffusion of Ge in i-AlPdMn was demonstrated by Frank et al. [114], who found a comparable Arrhenius law for Ge as for Al. The similarity of Si and Ge allows us to suggest that the diffusion of Si into i-AlPdMn should follow a similar Arrhenius behavior and also be vacancy-mediated. The increase of θ_D can then be understood in terms of quenching of vacancies in a near-surface region, resulting in the stabilization of the surface structure. The linearity of the rise of θ_D as a function of the amount of absorbed Si still awaits a satisfactory explanation.

In summary, in this Chapter we have presented the results of deposition of Si onto the pentagonal surface of i-AlPdMn. Structural investigations of the Si film by means of LEED and SEI show that Si grows in an amorphous structure. The θ_D of the pentagonal surface of i-AlPdMn rises with the adsorbed amount of Si. The observed increase of θ_D is 30 K for 60 Å incorporated Si indicating the stabilization of the surface structure. There is no adverse influence on the quasicrystalline order by the diffusion of Si into the bulk material. We suggest that Si diffusion into i-AlPdMn is a vacancy-mediated process. The θ_D of the freshly prepared surface is not affected by the amount of Si previously deposited onto the sample, implying that the ion sputtering and the thermal annealing of the surface restore the original chemical composition and the structure of the quasicrystalline surface.

Chapter 6

Conclusion and outlook

6.1 Summary

In this dissertation, new insights in the crystal-quasicrystal interfaces have been presented. The experimental investigations included AES, XPS, ARPES, LEED, and SEI. The main emphasis of this work was the influence of exposure of i-AlPdMn kept at high temperature to O_2.

Exposure of the clean fivefold-symmetry surface of i-AlPdMn kept at 700 − 800 K to O_2 results in the formation of a well-ordered oxide film. The remaining presence in the LEED patterns of diffraction spots arising from the quasicrystalline substrate indicates that the quasiperiodic long-range order at the surface is not disturbed by the oxidation process. These diffraction reflexes were observed even for exposure time corresponding to saturation (which was evaluated to be approximately 800 L). XPS investigations f the film show that only Al atoms are affected by the exposure to O_2, allowing us to argue that the film is exclusively an Al-ox film.

Similar diffraction patterns have been obtained after the oxidation at high temperature of (110) surfaces of CsCl alloys, as for instance NiAl. Films giving raise to these patterns have a thickness which was evaluated to be approximately 5 Å, are O terminated, and are extremely flat in the particular case of the oxidation of the NiAl(110) surface. Different models for the oxide film observed on NiAl(110) have been proposed in the last decades. All are based on a modified γ-Al-ox structure, which posses a giant unit cell with $\alpha = 88.7°$, $a_1 = 10.6$, and $a_2 = 17.9$ Å as lattice parameters. Discrepancies between the models occur in the exact position of the atoms in the unit cell. However, the maximal atomic displacement was measured to be approximately 0.3 Å, which has to be compared to the 100 Å coherence length of the LEED experiment. Contradicting stoichiometries of the films are present in the literature, until recently all the models proposed an Al_2O_3 suggest an $Al_{10}O_{13}$ stoichiometry.

The icosahedral structure has a strong affinity for the CsCl structure, where the (110) face of the latter share a same average structure as the fivefold-symmetry surface of i-AlPdMn. The Al-oxide observed on the quasicrystalline surface consists, as in case for NiAl(110), of modified γ-Al-ox domains with the same lattice parameters. Contrary to the oxidation of the binary alloy, five pairs of such domains rotated by 72° with respect to each other are observed, which is consistent with the fivefold symmetry of the substrate surface. Each pair is aligned along one of the five twofold-symmetry directions present in the pentagonal surface. The in-plane alignment of the Al-ox domains is a consequence of the strong relationship between the CsCl structure of sputtered-induced AlPd domains and the icosahedral structure of i-AlPdMn. The Al-ox domains have an average in-plane size of approximately 35 Å, as measured by means of LEED. The substrate long-range order is conserved after oxidation by the segregation of Al atoms to the surface, in order to re-establish the stoichiometry of the quasicrystalline phase. XPS investigations have been performed on the oxide films, however, the exact stoichiometry could not be determined due to the insufficient energy resolution of our system (x-ray source and energy analyzer).

We have also reported the results from the deposition of PbTe, CdTe, and Al onto the oxidized pentagonal surface of i-AlPdMn. Similar diffraction patterns were obtained in all three cases. They are characterized by the presence of diffraction rings instead of the more *common* spots. They correspond to films consisting of domains exhibiting random azimuthal orientations but well-defined polar orientations. They are (001) face for PbTe and (111) faces for CdTe and Al parallel to the fivefold-symmetry surface of i-AlPdMn. The average domain sizes for all three deposited materials are approximately 35 Å, which corresponds to the average domain size of the Al-ox domains. We argued that, in contrast to normal heteroepitaxy where the domain size of the deposited material is given by the lattice mismatch between the growing film and the substrate, it was here given by the size of the substrate. The PbTe film were observed to be air stable, which allowed us to perform ARPES measurements in an other UHV chamber. The results of theses investigations show that the Al-ox film isolates PbTe from the quasicrystalline surface, which is pre-required if one expects confinement effects in the semiconductor material domains. No dispersion of the electrons density of states in the PbTe film was observed in the photoemission spectra. We suggested that exposure to air changed the electrical properties of the PbTe film and therefore prevented us to observe confinement effects.

Confinement effects observed in Ag films deposited onto the fivefold-symmetry surface of i-AlPdMn and onto the tenfold-symmetry surface of d-AlCoNi were presented as model for confinement effects occurring due to the incompatible symmetries between crystalline films and quasicrystalline surfaces. Ag, in a fcc structure, exhibit in both cases its (111) face parallel to the substrates surface. By analyzing the Ag *sp*-derived QW states, we asserted that the interface with the quasiperiodic material constitutes an efficient barrier for electron propagation, due to lack of common point-group symmetries between Bloch-like and critical wave functions.

The results from the deposition of Si and Ge onto the fivefold-symmetry surface of i-AlPdMn have been, at last, repôrted. Investigations on the structure of the films, by means of LEED and SEI, have shown that both materials adopt, already for submonolayer coverage, a three-dimensional amorphous growth mode for a large substrate temperature range. At temperature above 370 K, AES and XPS measurements have shown that Si noticeably diffuses into the substrate, which prevents a probable crystalline growth mode at high temperature. We have then reported a linear increase of the surface θ_D as function of the amount of incorporated material. We suggested that Si diffusion into i-AlPdMn is a vacancy-mediated process. The θ_D of the freshly prepared surface is not affected by the amount of Si previously deposited onto the sample, implying that the ion sputtering and the thermal annealing of the surface restore the original chemical composition and the structure of the quasicrystalline surface.

6.2 Conclusion and outlook

Quasicrystal-crystal interfaces are fascinating systems. Due to the lack of exact information on the atomic positions at the quasicrystalline surfaces, models for the interface are almost impossible to construct. Nevertheless, recently some very interesting results from simulation of deposition of Al onto the tenfold-symmetry surface of d-AlCoNi were reported. Fourier transformation of the simulated films are in good agreement with the experimental diffraction patterns, opening another way to understand the complex mechanisms of the growth of crystalline material on quasiperiodic structures.

The formation of well-ordered Al-ox films onto i-AlPdMn is an exciting result. It gives the scientific community a complete new class of crystal-quasicrystal interfaces. Because the alignment of the oxide domains is a consequence of the strong affinity of the icosahedral structure for the CsCl structure, we now have, with the results of the deposition of ferromagnetic materials onto the same substrate, enough evidences that the optimum matching of the average structures plays a key role in the film orientations. Further theoretical works should be undertaken to gain a complete understanding of the growing mechanisms.

Controlled oxidation of quasicrystalline surfaces may open the door to applications of quasicrystals. First, the oxide was observed to be very stable when exposed to air and easily removable with ion sputtering without damaging the quasicrystalline bulk. Second, metal-oxide surfaces are known to be very good substrate for metal supported catalysis. The size of the Al-ox domains observed on i-AlPdMn could give the opportunity to study catalysis at a low nanometric scale. In this perspective, deposition of transition metals like Au or Pd onto the oxidized pentagonal surface of i-AlPdMn should be undertaken. The oxidation of quasicrystals make them bio-compatible and growth of active metal like Ag, which is used in the fight against cancer, could open promising perspective for quasicrystals in the field of bio-engineering. Recently, we were able to grow Co nanocrystallites on the oxide film. Their exact structure is still to be determined. We were also able to oxidize the tenfold-symmetry surface of d-AlCoNi. The oxide film undergoes a spectacular phase transition at approximately 900 K. Further investigations have to be performed to fully understand these later results.

Finally, confinement effects arising either from the incompatible symmetries, like for Ag on i-AlPdMn and d-AlCoNi, or from the size of the domains, like presumably in PbTe onto the oxidized fivefold-symmetry surface of i-AlPdMn, are of great interest for technical applications. Their observations give the possibility to use quasicrystal as template for self-assembly of electronic active devices in the nanometer range. Therefore, in-situ ARPES investigations of, for instance PbTe films, should be performed in order to reduce the contamination effect of air.

Bibliography

[1] Feynman, R. P. *Eng. Sci.* **23**, 22 (1960).

[2] Binnig, G., Quate, C. F., and Gerber, C. *Phys. Rev. Lett.* **56**, 930 (1986).

[3] Binnig, G. and Rohrer, H. *Ultramicroscopy* **11**, 157 (1983).

[4] Bohr, M. T. *IEEE Trans. Nanotech.* **1**, 1536 (2002).

[5] Moore, G. E. *Electronics* **38** (1965).

[6] Walter, C. *Scientific American* **August**, 32 (2005).

[7] Thompson, D. A. and Best, J. S. *IBM J. Res. Develop.* **44**, 311 (2000).

[8] Ito, T. and Okazaki, S. *Nature* **406**, 1027 (2000).

[9] Barth, J. V., Costantini, G., and Kern, K. *Nature* **437**, 671 (2005).

[10] Zhang, Z. and Lagally, M. G. *Science* **276**, 377 (1997).

[11] Zangwill, A., editor. *Physics at surfaces.* Cambridge University Press, (1990).

[12] Asryan, L. V. and Luryi, S. *J. Quantum El.* **37**, 905 (2001).

[13] Arakawa, Y. and Sakaki, H. *Appl. Phys. Lett.* **40**, 939 (1982).

[14] Vossmeyer, T., Katsikas, L., Giersig, M., Popovic, I. G., Diesner, K., Chemseddine, A., Eychmüller, A., and Weller, H. *J. Phys. Chem.* **98**, 7665 (1994).

[15] Colvin, V. L., Alivisatos, A. P., and Tobin, J. G. *Phys. Rev. Lett.* **66**, 2786 (1991).

[16] Goldstein, A. N., Echer, C. M., and Alivisatos, A. P. *Science* **256**, 1425 (1992).

[17] Costantini, G., Rastelli, A., Manzano, C., Songmuang, R., Schmidt, O. G., Kern, K., and von Känel, H. *Appl. Phys. Lett.* **85**, 5673 (2004).

[18] Hasegawa, Y., Kiyama, H., Xue, Q. K., and Sakurai, T. *Appl. Phys. Lett.* **72**, 2265 (1998).

[19] Daruka, I. *Phys. Rev. B* **66**, 132104 (2002).

[20] Shetchman, D., Blech, I., Gratias, D., and Cahn, J. W. *Phys. Rev. Lett.* **53**, 1951 (1984).

[21] Levine, D. and Steinhardt, P. J. *Phys. Rev. B* **34**, 596 (1986).

[22] Hippert, F. and Gratias, D., editors. *Lectures on Quasicrystals.* Les Editions de Pysique Les Ulis, (1994).

[23] Janot, C., editor. *Quasicrystals A Primer.* Oxford Science Publications, (1994).

[24] *Quasicrystal.* Encyclopedia Britannica.

[25] Trebin, H.-R., editor. *Quasicrystals: Structure and Physical Properties.* Wiley-VCH, (2003).

[26] Gödecke, T. and Lück, R. *Z. Metallkd.* **86**, 109 (1995).

[27] Hume-Rothery, W. *J. Inst. Met.* **35**, 295 (1926).

[28] Dubois, J.-M., Brunet, P., Costin, W., and Merstallinger, A. *J. Non-Cryst. Sol.* **334–335**, 475 (2004).

[29] Fournée, V., Ross, A. R., Lograsso, T. A., Evans, J. W., and Thiel, P. A. *Surf. Sci.* **537**, 5 (2003).

[30] Chernikov, M. A., Bianchi, A., and Ott, H. R. *Phys. Rev. B* **51**, 153 (1995).

[31] Saito, K., Ischioka, K., and Sugawara, S. *Phil. Mag.* **85**, 3629 (2005).

[32] Widjaja, E. J. and Marks, L. D. *Phil. Mag. Lett.* **83**, 47 (2003).

[33] Weisskopf, Y. *Growth of CsCl-type Domains on Icosahedral Quasicrystal Al-Pd-Mn.* Logos Verlag, Berlin, (2006).

[34] Weisskopf, Y., Burkardt, S., Erbudak, M., and Longchamp, J.-N. *Surf. Sci.* **601**, 544 (2007).

[35] Steurer, W. *Mat. Sci. Eng. A* **294**, 268 (2000).

[36] Steurer, W. and Haibach, T. *Acta Cryst.* **A55**, 48 (1999).

[37] Dmitrienko, V. E. and Astaf'ev, S. B. *Phys. Rev. Lett.* **75**, 1538 (1995).

[38] `http://www.physics.aamu.edu/Czlab/czc.php`.

[39] von Laue, M. *Materiewellen und ihre Interferenzen.* Akadem. Verl.-Ges. Becker and Erler, (1948).

[40] Kortan, A. R., Thiel, F. A., Chen, H. S., Tsai, A. P., Inoue, A., and Masumoto, T. *Phys. Rev. B* **40**, 9397 (1989).

[41] `http://en.wikipedia.org/wiki/Thermocouple`.

[42] Schmithüsen, F., Cappello, G., De Boissieu, M., Boudard, M., Comin, F., and Chevrier, J. *Surf. Sci.* **444**, 113 (2000).

[43] Ertl, G. and Küppers, J. *Low Energy Electrons and Surface Chemistry.* VCH Verlagsgesellschaft, Weinheim, (1985).

[44] Erbudak, M., Schulthess, T., and Welti, E. *Phys. Rev. B* **49**, 6316 (1994).

[45] Erbudak, M., Hochstrasser, M., Wetli, E., and Zurkirch, M. *Surf. Rev. Lett.* **4**, 179 (1997).

[46] `http://www.sbig.com`.

[47] Horn von Hoegen, M., Schmidt, T., Meyer, G., Winau, D., and Rieder, K. H. *Phys. Rev. B* **52**, 10764 (1995).

[48] Greber, T., Raetzo, O., Kreutz, T. J., Schwaller, P., Deichmann, W., Wetli, E., and Osterwalder, J. *Rev. Sci. Instrum.* **68**, 4549 (1997).

[49] Tanuma, S., Powell, C. J., and Penn, D. R. *Surf. Inter. Anal.* **17**, 911 (1991).

[50] King, D. A. and Woodruff, D. P., editors. *The Chemical Physics of Solid Surfaces and Heterogeneous Catalysis*, volume 1. Elsevier, Amsterdam, (1990).

[51] Franchy, R. *Surf. Sci. Rep.* **38**, 195 (2000).

[52] Bäumer, M. and Freund, H.-J. *Pogr. Surf. Sci.* **61**, 127 (1999).

[53] Goodman, D. W. *Surf. Rev. Lett.* **2**, 9 (1995).

[54] Freund, H.-J. *Angew. Chem. Int. Ed. Engl.* **36**, 452 (1997).

[55] Freund, H.-J. *Phys. Status Solidi (b)* **192**, 407 (1995).

[56] Erskine, J. L. and Strong, R. L. *Phys. Rev. B* **25**, 5547 (1982).

[57] Gassmann, P., Franchy, R., and Ibach, H. *Surf. Sci.* **319**, 95 (1994).

[58] Rose, V., Podgursky, V., Costina, I., Franchy, R., and Ibach, H. *Surf. Sci.* **577**, 139 (2005).

[59] Bardi, U., Atrei, A., and Rovida, G. *Surf. Sci. Lett.* **239**, 511 (1990).

[60] Libuda, J., Winkelmann, F., Bäumer, M., Freund, H.-J., Bertams, T., Neddermeyer, H., and Müller, K. *Surf. Sci.* **318**, 61 (1994).

[61] Stierle, A., Renner, F., Streitel, R., Dosch, H., Drube, W., and Cowie, B. C. *Science* **303**, 1652 (2004).

[62] Jäger, R. M., Kuhlenbeck, H., Freund, H.-J., Wuttig, M., Hoffmann, W., Franchy, R., and Ibach, H. *Surf. Sci.* **259**, 235 (1991).

[63] Kresse, G., Schmid, M., Napetschnig, E., Shishkin, M., Köhler, L., and Varga, P. *Science* **308**, 1440 (2005).

[64] Chang, S.-L., Chin, W. B., Zhang, C.-M., Jenks, C. J., and Thiel, P. A. *Surf. Sci.* **337**, 135 (1995).

[65] Popovic, D., Naumovic, D., Bovet, M., Koitzsch, C., Schlapbach, L., and Aebi, P. *Surf. Sci.* **492**, 294 (2001).

[66] Chang, S.-L., Anderegg, J. W., and Thiel, P. A. *J. Non-Cryst. Solids* **195**, 95 (1996).

[67] Schaub, T. M., Bürgler, D. E., Güntherodt, H.-J., Suck, J. B., and Audier, M. *Appl. Phys. A.* **61**, 491 (1995).

[68] Lüscher, R., Erbudak, M., Flückiger, T., and Weisskopf, Y. *App. Surf. Sci.* **233**, 129 (2004).

[69] Lay, T. T., Yoshitake, M., and Mebarki, B. *J. Vac. Sci. Technol. A* **20**, 2027 (2002).

[70] Lykhach, Y., Moroz, V., and Yoshitake, M. *App. Surf. Sci.* **241**, 250 (2005).

[71] Technical report, CrystalMaker Software Limited, Oxford University Begbroke Science Park Sandy Lane, Yarnton, Oxfordshire, OX5 1PF, UK.

[72] Wehner, B. I. and Köster, U. *Oxid. Metals* **54**, 445 (2000).

[73] Costina, I. and Franchy, R. *App. Phys. Lett.* **78**, 4139 (2001).

[74] Alivisatos, A. P. *Science* **271**, 933 (1996).

[75] Wise, F. W. *Acc. Chem. Res.* **33**, 773 (2000).

[76] Martiensen, W. and Warlimont, H., editors. *Springer Handbook of Condensed Matter and Materials Data.*

[77] Mackowski, S. *Thin Solid Films* **412**, 96 (2002).

[78] Schmitt-Rink, S., Miller, D. A. B., and Chemla, D. S. *Phys. Rev. B* **35**, 8113 (1987).

[79] Zúñiga-Pérez, J., Tean-Zaera, R., and Muñoz-Sanjosé, V. *J. Cryst. Growth* **270**, 309 (2004).

[80] Longchamp, J.-N., Burkardt, S., Erbudak, M., and Weisskopf, Y. ***Submitted***.

[81] Ferralis, N., Pussi, K., Finberg, S. E., Smerdon, J., Lindroos, M., McGrath, R., and Diehl, R. D. *Phys. Rev. B* **70**, 245407 (2004).

[82] Dashevsky, Z., Belechuk, A., Gartstein, E., and Shapoval, O. *Thin Solid Films* **461**, 256 (2004).

[83] Tatsuoka, H., Kuwabara, H., Nakanishi, Y., and Fujiyasu, H. *Thin Solid Films* **213**, 1 (1992).

[84] Luedtke, W. D. and Landman, U. *Phys. Rev. Lett.* **73**, 569 (1994).

[85] Kubiak, R. A. A., McGlashan, S. R. L., King, R. M., and Parker, E. H. C. *Appl. Phys. A* **40**, 7 (1986).

[86] Parker, E. H. C. and Williams, S. *Thin Solid Films* **35**, 373 (1976).

[87] Egerton, R. F. *Surf. Sci.* **24**, 647 (1971).

[88] Campbell, C. T. *Surf. Sci. Rep.* **27**, 1 (1994).

[89] Zhang, G. W., Stadnik, Z. M., Tsai, A.-P., and Ioue, A. *Phys. Lett.* **186**, 345 (1994).

[90] Fournée, V. and Thiel, P. A. *J. Phys. D: Appl. Phys.* **38**, 83 (2005).

[91] Zhou, S. Y., Gweon, G., Spataru, C. D., Graf, J., Lee, D.-H., Louie, S. G., and Lanzara, A. *Phys. Rev. B* **71**, 161403 (2005).

[92] Matsuda, I., Ohta, T., and Yeom, H. W. *Phys. Rev. B* **65**, 085327 (2002).

[93] Wu, Y. Z., Won, C. Y., Rotenberg, E., Zhao, H. W., Toyoma, F., Smith, N. V., and Qiu, Z. Q. *Phys. Rev. B* **66**, 245418 (2002).

[94] Fournée, V., Sharma, H. R., Shimoda, M., Tsai, A. P., Unal, B., Ross, A. R., Lograsso, T. A., and Thiel, P. A. *Phys. Rev. Lett.* **95**, 155504 (2005).

[95] Zhang, Z., Niu, Q., and Smith, C.-K. *Phys. Rev. Lett.* **80**, 5381 (1998).

[96] Davydov, D. N., Mayou, D., Berger, C., Gignoux, C., Neumann, A., Jansen, A. G. M., and Wyder, P. *Phys. Rev. Lett.* **77**, 3173 (1996).

[97] Chiang, T.-C. *Surf. Sci. Rep.* **39**, 181 (2000).

[98] Aballe, L., Rogero, C., Kratzer, P., Gokhale, S., and Horn, K. *Phys. Rev. Lett.* **87**, 156801 (2001).

[99] Lüscher, R. *Surface Science.* Nova Science New-York, (2005).

[100] Weisskopf, Y., Lüscher, R., and Erbudak, M. *Surf. Sci.* **578**, 35 (2005).

[101] Weisskopf, Y., Erbudak, M., Longchamp, J.-N., and Michlmayr, T. *Surf. Sci.* **600**, 2594 (2006).

[102] Ledieu, J., Unsworth, P., Lograsso, T. A., Ross, A. R., and McGrath, R. *Phys. Rev. B* **73**, 012204 (2006).

[103] Leung, L., Ledieu, J., Unsworth, P., Lograsso, T. A., Ross, A. R., and McGrath, R. *Surf. Sci.* **600**, 4752 (2007).

[104] Grigorovici, R. *Mat. Res. Bull.* **3**, 13 (1968).

[105] Richter, M. and Breitling, G. *Z. Naturf.* **13**, 988 (1958).

[106] Wagner, T. A. *Low temperature silicon epitaxy: Defects and electronic properties.* PhD thesis, (2003).

[107] Zurkirch, M., Crescenzi, M. D., Erbudak, M., and Hochstrasser, M. *Phys. Rev. B* **55**, 8808 (1997).

[108] Kittel, C. *Physique de l''etat solide.* Dunod, Paris, (1998).

[109] Lüscher, R., Flückiger, T., Erbudak, M., and Kortan, A. R. *Surf. Sci.* **532**, 8 (2003).

[110] Colella, R., Zhang, Y., Sutter, J. P., Ehrlich, S. N., and Kycia, S. W. *Phys. Rev. B* **63**, 014202 (2001).

[111] Lüscher, R., Flückiger, T., Erbudak, M., and Kortan, A. R. *Phys. Rev. B* **68**, 212203 (2003).

[112] Gierer, M., Van Hove, M. A., Goldman, A. I., Shen, Z., Chang, S.-L., Pinhero, P. J., Jenks, C. J., Anderegg, J. W., Zhang, C.-M., and Thiel, P. A. *Phys. Rev. B* **57**, 7628 (1998).

[113] Shen, Z., Stodlt, C. R., Jenks, C. J., Lograsso, T. A., and Thiel, P. A. *Phys. Rev. B* **60**, 14688 (1999).

[114] Frank, W., Blüher, R., and Schmid, I. *J. Alloys Comp.* **342**, 291 (2002).

Acronyms

d-	decagonal
i-	icosahedral
AES	Auger electron spectroscopy
LEED	low-energy electron diffraction
SEI	secondary-electron imaging
XPS	x-ray photoelectron spectroscopy
ARPES	angle-resolved photoemission spectroscopy
UHV	ultra-high vacuum
QW	quantum-well
QD	quantum-dot
Al-ox	aluminum-oxide

Publications

Crystalline textures on the Al-Ni-Co quasicrystal surface
M. Erbudak, J.-N. Longchamp, Y. Weisskopf
Turk. J. Phys. 29 (2005) 277

In situ formation of a new Al-Pd-Mn-Si quasicrystalline phase on the pentagonal surface of the Al-Pd-Mn quasicrystal
J.-N. Longchamp, M. Erbudak, Y. Weisskopf
J. Phys. IV France 132 (2006) 117

Ni deposition on the pentagonal surface of an icosahedral Al-Pd-Mn quasicrystal
Y. Weisskopf, M. Erbudak, J.-N. Longchamp, T. Michlmayr
Surf. Sci. 600 (2006) 2594

Self-assembled nano-structures on the icosahedral Al-Pd-Mn quasicrystal
M. Erbudak, J.-N. Longchamp, Y. Weisskopf
Turk. J. Phys. 30 (2006) 213

Quantum size effects arising from incompatible point-group symmetries
P. Moras, Y. Weisskopf, J.-N. Longchamp, M. Erbudak, P.H. Zhou, L. Ferrari, C. Carbone
Phys. Rev. B 74 (2006) 121405

The quasicrystal-crystal interface between icosahedral Al-Pd-Mn and deposited Co: evidence for the affinity of the quasicrystal structure to the CsCl structure
Y. Weisskopf, S. Burkardt, M. Erbudak, J.-N. Longchamp
Surf. Sci. 601 (2007) 544

Stabilization of the pentagonal surface of the icosahedral AlPdMn quasicrystal by controlled Si absorption
J.-N. Longchamp, M. Erbudak, Y. Weisskopf
Appl. Surf. Sci. 253 (2007) 5947

Formation of a well-ordered ultra-thin Aluminum-oxide film on icosahedral AlPdMn quasicrystal
J.-N. Longchamp, S. Burkardt, M. Erbudak, Y. Weisskopf
Phys. Rev. B (accepted)

CdTe and PbTe nanostructures on the oxidized pentagonal surface of an icosahedral AlPdMn quasicrystal
J.-N. Longchamp, S. Burkardt, M. Erbudak, Y. Weisskopf
Surf. Sci. (accepted)

Curriculum Vitae

Last and first name	Longchamp Jean-Nicolas
Date of birth	January 22, 1980
Citizen of	L'Isle, VD, Switzerland

Education

1984 – 1986	École enfantine Lausanne
1986 – 1990	École primaire Lausanne
1990 – 1995	École secondaire Lausanne, section C
1995 – 1998	Gymnase de la Cité Lausanne, Maturité type C
1998 – 1999	Studies of medicine at the University of Lausanne
1999 – 2004	Studies of physics at the Swiss Federal Institute of Technology Zurich
04/2004 – 09/2004	Diploma thesis on *Cloning and expression on the DNA repair and cell-cycle associated protein Mus81* in the group of Dr. A. Bergkvist, Institute for Molecular Biology, University of Göteborg, Sweden
10/2004 – 09/2007	PhD student in the research area of *Thin films on icosahedral AlPdMn quasicrystal* in the group of Prof. Dr. M. Erbudak, Institute for Solid State Physics, Department of Physics, Swiss Federal Institute of Technology Zurich

Thanks

A lot of people have contributed to the success of this dissertation. I would like here to thank them all. In particular, I would like to express my infinite gratitude to my parents and my whole family for their support not only during my dissertation but during all my life. I would like to thank also specially Prof. Dr. Mehmet Erbudak for his patience and innumerable help in the lab and for the tremendous time I had working with him; Prof. Dr. Danilo Pescia for his support and his interest in this work; Dr. Rouven Lüscher, Sven Burkardt, Dr. Thomas Michlmayr, Dr. Andreas Vaterlaus, Dr. Urs Ramsperger, Niculin Saratz and specially Dr. Yves Weisskopf for their assistance and the pleasure working together; Dr. Paolo Moras for the good collaboration and the good time in Trieste; Prof. Dr. Walter Steurer for his interest in this dissertation; and last but not least, Eliane Morf for her incredible support during the last months.